Manual of Ready-Mixed Concrete

J.D. DEWAR
Director
British Ready Mixed Concrete Association
and
R. ANDERSON
Product Officer
British Aggregate Construction Materials Industries

BLACKIE ACADEMIC & PROFESSIONAL
An Imprint of Chapman & Hall
London · Glasgow · New York · Tokyo · Melbourne · Madras

**Published by Blackie Academic & Professional, an imprint of Chapman & Hall,
Wester Cleddens Road, Bishopbriggs, Glasgow G64 2NZ, UK**

Chapman & Hall, 2–6 Boundary Row, London SE1 8HN, UK

Blackie Academic & Professional, Wester Cleddens Road, Bishopbriggs,
Glasgow G64 2NZ, UK

Van Nostrand Reinhold Inc., 115 Fifth Avenue, New York NY10003, USA

Chapman & Hall Japan, Thomson Publishing Japan, Hirakawacho Nemoto
Building, 6F, 1-7-11 Hirakawa-cho, Chiyoda-ku, Tokyo 102, Japan

DA Book (Aust.) Pty Ltd, 648 Whitehorse Road, Mitcham 3132, Victoria,
Australia

Chapman & Hall India, R. Seshadri, 32 Second Main Road, CIT East, Madras
600 035, India

First edition 1988
Second edition 1992

£47,98ir

© 1992 Chapman & Hall

Typeset in 10/12 pt Times Roman by Thomson Press (India) Ltd, New Delhi
Printed in Great Britain by St. Edmundsbury Press, Bury St. Edmunds,
Suffolk

ISBN   0 7514 0079 3   0 442 30866 3

A catalogue record for this book is available from the British Library

Library of Congress Cataloging-in-Publication data available

666.893

069755

∞ Printed on permanent acid-free text paper, manufactured in accordance
with the proposed ANSI/NISO Z 39.48-199X and ANSI Z 39.48-1984

Preface

The first edition of this manual was warmly received as a straightforward and well-written guide to the technology and practice of the ready-mixed concrete industry. The industry is constantly changing, not only by a self-driven desire to improve standards and quality, but also in response to the changing needs of the marketplace, and the requirements to adapt to new and revised codes of practice.

The manual has been completely updated to take account of changes in British Standards—in particular those relating to concrete specification and testing, and to cements. A particularly important inclusion is the concept of the Designated Mix, which has been welcomed by the construction industry at large as a major improvement to the way of specifying concrete for particular end-uses. Throughout, we have also taken the opportunity to amend and add to the text where we have perceived a need to present information and ideas differently. As a result this revised edition is an improvement on the first edition and should serve the needs of the market equally well.

We have retained the overall structure and format of the successful first edition and, as before, to enable clear and logical presentation of the information, the book is divided into two main sections: Part 1, Technology, and Part 2, Practice. The Technology section provides the background to the technical aspects of materials, concrete, control and testing, with the Practice section dealing with methods of concrete production, specification, construction, and quality assurance.

As before, the book is aimed at a wide readership within the construction industry. It is equally relevant to the contracts manager, the site engineer, the buyer working with a contractor, and will be of considerable value to the consulting engineer, the architect and the quantity surveyor. It will be an invaluable handbook for materials suppliers and testhouses and for all who supply plant, equipment or materials, or who make, sell test or use ready-mixed concrete. Lecturers and researchers in universities and colleges will find it a useful source of reference.

JDD
RA

Acknowledgements

The authors gratefully acknowledge that many of the principles and methods described are the distillation of years of experience generated by the many individuals who collectively make up the ready-mixed concrete industry, its suppliers and its clients. The net result may be taken as a summation of views for which the authors are privileged to be the selectors and arbiters. Naturally, the authors take responsibility for expressing the views and the facts and for any omissions or errors.

To minimize the risk of error, Mr B.V. Brown, Senior Technical Executive of Ready Mixed Concrete (United Kingdom) Ltd, kindly read and commented upon the whole of Part 1, and the authors are particularly grateful to him. On two specialized subjects—cements and admixtures—thanks are due for expert comments from Mr G.F. Masson, National Technical Manager, Blue Circle Industries plc, and Dr P.C. Hewlett, Managing Director, Cementation Research Ltd and visiting professor at the University of Dundee. For providing particular help with recent developments we wish to thank Mr P.M. Barber (QSRMC) and Dr T.A. Harrison (BCA).

The comments and criticism of Mr P.E.D. Howes of ARC Ltd at the drafting stage were much appreciated by both the authors, and material provided by Mr P. Male, then of Steetley Quarry Products, Mr D. Bickley, then of Pioneer Concrete (UK) Ltd., and Mr P.N. Staples of Tilcon Ltd., provided the basis for much of Part 2.

The following published papers have been used by permission of their authors and publishers in suitably modified, updated or abbreviated forms: 'BRMCA Guide: Concrete Mixes, an Introduction to the BRMCA Method of Mix Design', and 'BRMCA Guide: BRMCA Concrete Control System' (BRMCA); 'Monitoring concrete by the CUSUM system' (B.V. Brown and the Concrete Society); 'Ready Mixed Concrete Mix Design' (*Municipal Engineer*); *Testing Concrete for Durability* (Palladian Publications); 'Quality Scheme for Ready Mixed Concrete Technical Regulations' (QRSMC); 'The workability of ready-mixed concrete' (RILEM).

A number of tables and figures have been adapted from published papers as acknowledged in the references. Particular thanks are due to the following for permission to reproduce data, tables or figures: Mr F.W. Beaufait; Mr B.V. Brown; Professor R.K. Dhir; Dr A.M. Neville; Mr K. Newman; Mr B. Osbaeck; Professor S. Popovics; Mr R. Ryle; Mr R.E. Spears; ACI; BACMI; BRMCA; BSI; BCA; Concrete International; The Concrete Society; Controller of HM Stationery Office; ERMCO; *Municipal Engineer*; *New Civil Engineer*; Palladian Publications; Pitman Books Ltd; Pergamon Press Ltd; RILEM; RMC Technical Services Ltd.

Extracts from BS 6089:1981 are reproduced by permission of BSI. Complete copies can be obtained from them at Linford Wood, Milton Keynes MK 14 6LE.

Thanks are given to Mr Robert Phillipson, Director General of BACMI, for permitting the text to be transferred to a BACMI word processor which has made the editing and re-editing of drafts a far simpler task than it would have been otherwise. We also thank the publishers for their support.

Contents

PART 2: PRACTICE

Abbreviations

A/C	Aggregate/cement ratio
ASR	Alkali–silica reaction
BACMI	British Aggregate Construction Materials Industries
BCA	British Cement Association
BD	Bulk density
BRE	Building Research Establishment
BRMCA	British Ready Mixed Concrete Association
BS	British Standard
BSI	British Standards Institution
C	Characteristic compressive strength, as in grade C40
C&CA	The Cement and Concrete Association (now British Cement Association)
CEGB	Central Electricity Generating Board
CMF	Cement Makers' Federation (now British Cement Association)
CP	Code of Practice
CSTR	Concrete Society Technical Report
CTMA	Construction Testing Manufacturers Association
ggbs	Ground granulated blastfurnace slag
kg	Kilogram
kg/m^3	Kilograms per cubic metre
km	Kilometre
lasrpc	Low alkali sulphate-resisting Portland cement
mm	Millimetre
NACCB	National Accreditation Council for Certification Bodies
NAMAS	National Measurement Accreditation Service
N/mm^2	Newtons per square millimetre
pc	Portland cement
pfa	Pulverized-fuel ash
QA	Quality Assurance
QC	Quality Control
QSRMC	Quality Scheme for Ready Mixed Concrete
RD	Relative density
s.d.	Standard deviation
srpc	Sulphate-resisting Portland cement
SSD	Saturated and surface dry
TMS	Target mean strength
W/C	Water/cement ratio

Part 1
TECHNOLOGY

Photograph courtesy RMC (UK) Ltd.

Introduction: History of ready-mixed concrete

The history of concrete in Britain [1, 2, 3] dates back to Roman times, but it was not until the 1930s that concrete was supplied ready-mixed in the UK. The constituent materials come from the earth in bulk, so concrete is most economically produced by handling the materials in bulk and mixing them in bulk. It was the advantages of scale and efficiency of mechanical mixing, plus improved control resulting from weighing the ingredients, that opened the way for the supply of concrete ready-mixed.

Ready-mixed concrete was patented in Germany in 1903, but the means of transporting it had not developed sufficiently well to enable the concept to be exploited. There were significant developments in the USA in the first quarter of the 20th century: the first delivery of ready-mixed concrete was made in Baltimore in 1913, and the transit-mixer was born in 1926.

In the UK it was not until 1930 that a Dane, K.O. Ammentorp [1], who had been involved in starting ready-mixed concrete in Copenhagen, emigrated to England, and in 1931 erected a plant at Bedfont, west of London, near what was to be eventually the site of Heathrow Airport. He suffered from planning delays, even in those days! The company, Ready Mixed Concrete Ltd, operated the plant housing a 2 cu yd central mixer, supplying six $1\frac{3}{4}$ cu yd capacity agitators (Fig. I.1), with an output of 40 cu yd/h. Aggregates were contained in a four-compartment bin of about 100 cu yd capacity. The cement was man-handled in bags.

At about the same time, the British Steel Piling Company became interested in transit-mixers and imported two from the USA. These had a mixing capacity of about 5 cu yd and were filled (with difficulty!) through a small hole in the back.

The next company to start was the Scientific Controlled Concrete Co Ltd of Staines, in 1934. They used Jaeger truckmixers, produced under licence by Ransome & Rapier Ltd. It appears that this firm operated for only a short time; some of their equipment was taken over by Truck Mixed Concrete (Southampton) Ltd and used in the Winchester Bypass. Next came Jaeger System Concrete Ltd, Glasgow, which much later was taken over by Tilcon Ltd. In 1936 the Express Supply Concrete Ltd was founded as a subsidiary of Balfour Beatty Ltd. It had two plants, the first at Paddington and the second at Alperton, with a total of 30 Jaeger truckmixers. At these plants the cement was delivered in metal 5-ton capacity bulk containers, which were lifted off the

Figure I.1   Early towerplant with truckmixers and agitator. Courtesy of Tilcon Ltd

delivery lorries and emptied into bins by opening a gate at the bottom of the
container. A pump was then used to elevate the cement from the ground bins
to the bins over the cement weighing floor; a similar system is used for handling
large plastic bags of bulk cement in some locations today. Another early
producer from the aggregate side of the business was Trent Gravel Ltd, whose
first plant was erected at Attenborough, near Nottingham, in 1939. At the
outbreak of World War II in 1939, there were only six firms producing ready-
mixed concrete in England and Scotland, one supplying concrete in agitators
from a central mixer and the others using the truckmixer system.

The handling of materials in bulk has dominated the design of ready-mixed concrete plants from the start. They were aligned initially to quarry practice, using drag-line excavators for aggregates and bucket elevators for cement. Long, high conveyors were used to elevate the aggregates to a height that allowed them to be stored above the weigh-scales and to be gravity-fed as required, from storage bin to weigh-hopper, from weigh-hopper to mixer and from mixer to the delivery vehicle. These tower plants (Fig. I.2) came into vogue in the late 1930s and after World War II. Some of them saw wartime service on airfield construction and other projects requiring large volumes of concrete. Tall plants were more acceptable in quarries and gravel pits but, with the requirement to provide plants nearer the markets for the concrete, plant designers had to provide units that were more environmentally acceptable.

The development of chevron conveyor belts meant that steeper angles could be used in elevating the aggregates, thus requiring less space. Separating the storage and weighing of aggregates from the storage and weighing of the cement meant that the height of plants could be considerably reduced. The split-level plants introduced in the late 1960s are still popular today and now dominate the industry, but the demands of planning authorities for lower-profile plants have led to the production of some even more compact plants. With compactness, plant transportability became viable, and now some ready-mixed concrete producers provide plants on major construction projects.

Ready-mixed concrete is a service as well as a product, and delivery techniques have developed along with the production side. In the 1930s,

Figure I.2   Conical agitator (1950s). Courtesy of C & CA and RMC Ltd

conical agitators and Jaeger truckmixers did yeoman service, although they were susceptible to the rear door opening in transit, resulting in premature discharge. Their use continued into the early 1950s, and as late as 1985 a site foreman in Central Scotland rang up the shipping office to ask for 'a Jaeger of concrete'! The industry has generally encouraged the collection of concrete from its central mixer plants. Normal haulage lorries were frequent callers in the early days, then came tippers. Some companies operated telehoist tippers for a while, but it was the rotating drum mixer which prevailed and developed into the modern truckmixer which has really made the supply of ready-mixed concrete commercially viable.

The current shape of truckmixer drum, relying on reversible action for loading and mixing then contra-rotating for discharging, developed into the late 1950s, so that by the 1980s the most common size, based on a three-axle chassis, mixes and transports six cubic metres of concrete. Larger units (carrying nearly $9\,m^3$) and smaller ones (down to $2.5\,m^3$) are in use, but the industry has accepted the $6\,m^3$ truckmixer as the basic fleet unit.

The ready-mixed concrete industry has tended to outpace the construction industry in its willingness to innovate. Not all new ideas survived commercial pressures, mainly because purchasers were not always willing to pay a premium for an improvement in quality or service. Wet hoppers were supplied to sites into which the truckmixers could quickly discharge the concrete, allowing the contractor to collect and place the concrete at his own pace. Site conveyors were tried, and there are a few truckmixers operating with conveyors mounted on them. The early versions were heavy, resulting in a considerable reduction in the volume of concrete that could be carried in the drum. Developments using aluminium to lighten the conveyor have revived their popularity.

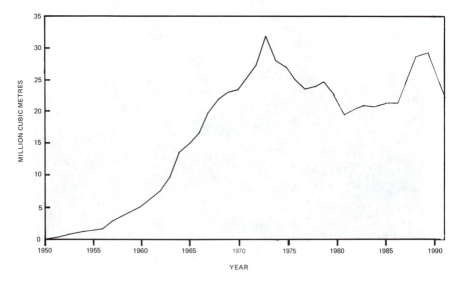

Figure I.3   Ready-mixed concrete in Britain. Courtesy of BRMCA and BACMI

There are even truckmixers with concrete pumps mounted on the chassis, but although such innovation is appreciated abroad, it has not gained favour in the UK. The ready-mixed concrete industry entered the concrete pumping business with enthusiasm, but lack of response from customers meant that the pumps were eventually sold off to a few concrete pumping specialists. In Britain, the percentage of ready-mixed concrete that is pumped is markedly lower than in the rest of Europe.

Cement weighing recorders were in vogue in the late 1960s, but the electro-mechanical gear did not stand up well to the dust-laden atmosphere, and 20 years had to elapse before solid-state electronics arrived to allow the manufacture of reliable processing, batching and recording equipment at economic prices. The processing systems have evolved from lever-arms, to servo-systems using compressed air, hydraulics and electronics, to computers and micro-processors, each step aimed at giving improved control of the whole production process.

The acceptance and growth of the ready-mixed concrete industry between the 1950s and 1974 was quite remarkable, with 31 million cubic metres per year being produced at the peak. The downturn in the construction industry was naturally reflected in the concrete production figures, and a period of rationalization and consolidation followed (Fig. I.3).

The short history of the ready-mixed industry has been dominated by the need to produce and deliver a high-quality product economically. The fact that most *in-situ* concrete is now supplied ready-mixed is a measure of how successful the industry has been, in terms of quality, service, and price to the customer.

Seeking to provide assurance of the quality of ready-mixed concrete has always been a requirement of the industry and was first manifested in the BRMCA Authorization Scheme, launched in 1968. This scheme introduced and enforced standards ensuring that each registered plant was able to produce concrete of the requisite quality and quantity. An additional section was added to the scheme in 1972, providing a very early approach to QA, by introducing auditing and certification of quality control procedures. In 1982, when BACMI was formed, it introduced a code of practice based on BS 5750 quality systems.

The Quality Scheme for Ready Mixed Concrete (QSRMC) was established in 1984 and was granted the ninth certificate accredited by the NACCB. With its independent Governing Board regulations based on BS 5750: Part 1—Quality systems, BS 5328—Concrete and NAMAS regulations for laboratory performance, the scheme epitomizes the current philosophy of third party QA.

The cost to the industry of providing ever-increasing levels of quality assurance has been considerable, while the complexity of mixes being specified has increased further the costs of quality control. The fact that most in-situ concrete is now supplied ready-mixed is a measure of how successful the industry has been in terms of demands for quality, service, and price.

1 Materials for concrete

Ready-mixed concrete is a composite material which, like all well-designed composites, has resultant properties that combine the best qualities of the component materials.

As the number of components is increased, so the range of properties and uses are increased, with greater opportunity for optimization and economic benefit. Fig. 1.1 illustrates the benefits obtained in reducing water content as the number of solid components is increased from one for cement paste, to two for mortar to three for concrete.

Typical concrete consists of only 4–6 components:

Portland cement
Sand
Coarse aggregate (one, two or three fractions)
Water.

The options and number of components can be increased by adopting various combinations of the following:

Different types of Portland cement
Ggbs or pfa

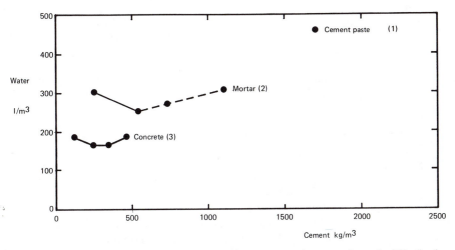

Figure 1.1 Influence of the number of solid components on the water demand of Portland cement-based composites. Redrawn and adapted from Dewar [12]

Natural sand, crushed rock fines or fine lightweight or heavyweight aggregate

Different maximum sizes of natural gravel, crushed rock or blastfurnace slag as coarse aggregates

Light- or heavyweight coarse aggregates

Admixtures.

The opportunities for optimization do not end with the choice of materials. There is also considerable scope for selection of proportions appropriate to specification requirements and the intended uses of the concrete.

1.1 Aggregates

Aggregates have two prime functions in concrete:

(i) Providing concrete with a rigid skeletal structure
(ii) Reducing the void space to be filled by the cement paste.

By selecting different sizes and types of aggregates and different ratios of aggregate to cement, a wide range of concretes can be produced economically to suit different requirements.

Important properties of an aggregate which affect compliance with British Standards and affect the performance of ready-mixed concrete are [13]:

Nominal maximum size
Grading and mean size
Silt, clay or fine dust content
Shape and surface texture
Water absorption
Relative density
Bulk density
Moisture content.

Other properties which have particular importance for some aggregates or for some special uses are:

Chloride content
Susceptibility to alkali-silica reaction
Deleterious materials content
Moisture movement.

For use generally as an aggregate, materials need to be essentially chemically inert and physically strong and stable. Most natural rocks, whether massive or broken down by nature into gravel and sand, make first-class concreting aggregates [5]. The rock types in most common use are flint, quartzite, limestone, gritstone, granite and basalt. A more complete list is shown in Table 1.1.

Table 1.1 Rock types commonly used as concreting aggregates. Alphabetic list selected from BS 812: Part 102: 1989, Table 2

andesite	dolomite	hornfels
basalt	flint	limestone
chert	gabbro	quartzite
diorite	granite	sandstone
dolerite	gritstone	syenite

Within any rock type, the range of concrete performance may be quite wide. It is thus important to make comparisons on the basis of concrete performance rather than to assume similarity on the basis of rock type. Limestone, for example, covers a wide range of materials from Jurassic oolitic limestone to dense Carboniferous limestone; both behave differently, but both can be used satisfactorily in properly designed concretes [6].

Most gravels contain a mixture of rock types, but one type may predominate, e.g. in the south of England gravels are mostly flint, but may contain some quartzite, gritstone and possibly limestone.

Fine aggregates used in ready-mixed concrete are predominantly natural silica sands, but there are some in which other materials predominate, e.g. limestone. There is some use of crushed rock fines [4], but usually only in combination with a sand.

Considerable use is made of crushed rock for coarse aggregate. But uncrushed or crushed gravel is the predominant material, as shown by Table 1.2. There has been considerable growth in the use of marine gravels and sands to supplement land materials [7], [8]. Blastfurnace slag is the only artificial dense aggregate commonly used. Lightweight, and less commonly heavyweight, aggregates are used for special types of concrete.

1.1.1 *Maximum aggregate size*

The *nominal maximum aggregate size* is specified by the concrete user to take account of the section thickness and steel reinforcement spacing in the

Table 1.2 Production of aggregates for concrete in the UK [9]

Aggregates	Million tonnes (1990)
Crushed rock	18.8
Gravel	
Land-won	27.2
Marine-dredged	9.3
Sand	
Land-won	31.2
Marine-dredged	6.0

concrete construction. A small proportion of oversize and undersize material is permitted in order not to unduly restrict the way in which aggregate is screened during its production. The normal maximum sizes are 40 mm, 20 mm and 10 mm. Use of a larger maximum size extends the range of sizes of particles in concrete and permits lower fine aggregate content and lower water content.

1.1.2 *Grading*

The distribution of the sizes of aggregate particles is called the *grading*. Grading is usually described in terms of the cumulative percentage by mass of aggregate passing particular sieves, the commonest of which have apertures which are approximately twice the size of the next sieve below in the series:

50 mm
37.5
20
10
5
2.36
1.18
0.6 (600 microns)
0.3 (300 microns)
0.15 (150 microns)

For historic reasons, the 5 mm sieve has become the size separating coarse from fine aggregate, although some small fraction of a coarse aggregate is permitted to pass 5 mm and some fine aggregate is permitted to be coarser than 5 mm.

Coarse aggregates are described as *graded* (i.e. having more than one size of particle), or *single-sized* (mainly retained between two adjacent sieves in the upper part of the list). Typical 20 mm graded and 20 mm single-sized aggregates complying with BS 882 might have gradings similar to those shown in Table 1.3. In the examples, for the 20 mm–5 mm graded material, 90% lies between the two sizes used to describe it and, for the 20 mm single-sized

Table 1.3 Typical gradings for 20 mm graded and single-sized coarse aggregates.

Sieve size (mm)	Percent cumulative passing by mass	
	20 mm–5 mm graded	20 mm single-sized
37.5	100	100
20	95 ⎫	92 ⎫
10	40 ⎬ (90%)	12 ⎬ (80%)
5	5 ⎭	2
2.36	—	—

material, 80% lies between the sieve used to describe it and the next size below. These gradings are merely examples; different values will be obtained for other sources and they may vary with time and from consignment to consignment.

For compliance with BS 882, the standard for concreting aggregates, the permitted ranges for the two materials are as shown in Table 1.4.

Table 1.4 BS 882 limits for 20 mm graded and single-sized coarse aggregates.

| Sieve size (mm) | BS 882 limits (% cumulative passing) | |
	20 mm–5 mm graded	20 mm single-sized
37.5	100	100
20	90–100	85–100
10	30–60	0–25
5	0–10	0–5

Fine aggregates complying with BS 882 are now divided into three categories (coarse, medium and fine), whereas until recently there were four categories (zones 1, 2, 3 and 4). A typical medium grading might be as shown in Table 1.5.

Table 1.5 Typical medium grading of fine aggregate to BS 882.

Sieve size	% cumulative passing by mass
10 mm	100
5	95
2.36	80
1.18	70
600 microns	50
300	25
150	10

Table 1.6 Comparison of new and obsolete fine aggregate grading limits of BS 882.

| System | BS 882 limits for the 600-micron sieve (percentage passing) | | |
	C	M	F
New	15–54	25–80	55–100
	Zone 1	Zone 2	Zone 3	Zone 4
Obsolete	15–34	35–59	60–79	80–100

A comparison is given in Table 1.6 of the ranges permitted by BS 882 for the percentage passing the 600-micron sieve for the new and obsolete systems of fine aggregate grading classification. The new system permits more sands to be accepted which have been found satisfactory for use in concrete. Fine aggregates in each of these grading categories can be used for making high quality concrete; however, a sand at the coarse end of category C is usually less likely to be suitable for concretes of low cement content and therefore less suitable for general purpose concreting. Some very fine sands, e.g. 90% retained between 300- and 150-micron sieves, which are too fine for the F category of BS 882, can be used to produce excellent concrete [10, 11], which underlines the still arbitrary nature of current classification.

An *all-in* aggregate is an aggregate already blended for use in concrete, containing coarse and fine aggregate either processed as one material or reconstituted by blending of separately processed materials. All-in aggregate is normally permitted only for less critical concrete construction.

The overall grading of the total aggregate is either a *continuous* or a *gap grading*. A continuous grading implies some proportion between each sieve whereas a gap grading will be missing intermediate material. A gap in the overall aggregate grading commonly occurs when fine sands are used. Both gap and continuous gradings are equally satisfactory for use in concrete.

Gradings of aggregates are often shown as grading curves or charts. Fig. 1.2 shows gradings for single-sized 20 and 10 mm coarse aggregates, and for a medium sand together with the overall grading for the three aggregates combined in the proportions 40/20/40. The bottom scale for sieve size is a logarithmic scale, the scale distance between adjacent sizes being approximately equal. The same materials are detailed in Table 1.7 together with the proportioning to show how the overall grading can be calculated. The difference between adjacent percentages in the final column is the percentage

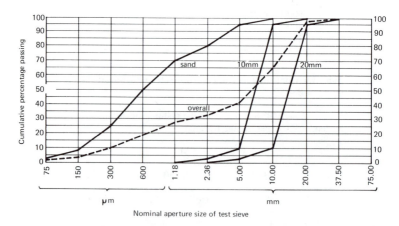

Figure 1.2   Examples of gradings of aggregates

Table 1.7 Example of individual and combined gradings.

Sieve (mm)	Grading (% passing by mass)			Proportioning*			Overall grading**
	20 mm SS	10 mm SS	sand	40% × 20 mm	20% × 10 mm	40% × sand	
37.5	100	100	100	40	20	40	100
20	95	100	100	38	20	40	98
10	10	95	100	4	19	40	63 ⎫
5	2	10	95	1	2	38	41 ⎬ 22%
2.36		2	80		0	32	32 ⎭
1.18			70			28	28
0.6			50			20	20
0.3			25			10	10
0.15			7			3	3

*% of material × % passing/100
**Sum of previous three columns.

material between the corresponding sieves, e.g. 22% lies between the 10 and 5 mm sieves. It will be apparent from the final column in the table and the resultant curve in Fig. 1.2 that the overall aggregate grading is continuous, not gapped.

Gradings of fine and coarse aggregates are important properties of aggregates for concrete because of their influence on packing, and thus voidage, which will in turn influence the water demand and cement content of concrete [12]. There are no recognized ideal gradings for aggregates in the United Kingdom. Uniformity of grading within and between consignments is vital [13]. Knowledge of the current grading of an aggregate is important, so that allowance can be made for the effects of any changes. There are rare instances of types of construction or finish which may require special restrictions of grading, e.g. some road-paving finishing operations achieve better results when 10 mm aggregate is omitted; some exposed aggregate finishes may require omission of 10 mm aggregate or the use of fine sand for architectural reasons.

The proportion of fine to coarse aggregate is selected to suit the materials, their proportioning and the type and use of concrete. This is considered further under mix design (Chapter 4).

1.1.2.1 *Mean size.* There is no standard way of estimating mean size of a graded material. One simple way [12] is to estimate, from the grading, the sieve size at which 50% of material is retained or passed. In the case of the gradings in Table 1.7, the 20 mm, 10 mm and fine aggregates have mean sizes of about 15 mm, 7 mm and 0.6 mm.

A small mean size for the fine aggregate relative to that of the coarse aggregate is important to minimize particle interference, which will affect water demand and workability of concrete adversely. On the other hand, the mean size of the fine aggregate must not be so small that it results in interference with cement grains.

1.1.3 *Silt, clay and fine dust*

The finest particles in fine aggregate are essential constituents needed to supplement cement to reduce the void content, particularly in lean concretes.

They fulfil a vital function by providing cohesion, in aiding the pumpability of concrete and in reducing the tendency for water to bleed from concrete. It is important that over-emphasis on cleanliness of aggregates does not lead to a loss of these vital ingredients. In the case of marine aggregates, it is of particular importance to ensure that loss of fine material is minimized during dredging.

However, too much fine material, particularly in concretes of high cement content, is less desirable. BS 882 places a maximum limit of 3% for *silt and clay* content of all natural sands, irrespective of use, where silt and clay are defined as material passing a 75-micron sieve. Similarly, a corresponding limit of 1% is placed on silt, clay or fine dust content of coarse aggregate.

It is stressed that these limits are arbitrary and the case can be made for higher limits in specific cases [14]. Indeed, lean mixes can tolerate and actually need a higher proportion of silt or dust to assist cohesion. Cement is a very expensive substitute for silt!

Crushed rock fines have a higher void content than natural sand of a similar grading, and therefore need a higher content of material passing the 75-micron sieve, which is one reason why BS 882 allows up to 15% *fine dust* compared with only 3% for silt and clay in sands for general-purpose concreting.

Fine dust or clay coating could adversely affect bond with cement paste. It is sometimes observed that oven-dried aggregates prepared in the laboratory for testing are dulled by the presence of silt, clay or dust. This should not be considered detrimental unless it is excessive, because fine particles in suspension in water will naturally appear as coating when the aggregate has been dried. Clay lumps of significant size and number are obviously unwanted.

The proportion of certain swelling clay minerals such as montmorillonite, in sands, may need to be reduced to avoid high water demands and excessive moisture movement of concrete [5], [15].

The common field settling test for clay and silt by settlement of sand in a saline solution in a measuring cylinder is a very simple and useful test, but is misleading because the fine material settles in a very loosely packed layer on top of the coarser particles. As much as 10%, or even more, of silt measured by bulk volume in this test may not represent much more than 3% when measured by weight. Because of this, BS 882 permits up to 10% in the volume test before it is necessary to perform the more tedious but precise test by weight.

1.1.4 *Shape and texture*

Shape and texture of aggregates are affected by the nature and geological history of the materials and by the aggregate production process, more especially the degree and method of crushing.

Generally, round and smooth particles, particularly in fine aggregate, will aid workability of concrete and lead to a lower voidage and lower water demand and, incidentally, to less plant wear. Angular and rough particles of coarse aggregate will bond better to cement paste and often lead to higher strengths in mixes of high cement content.

A proportion of flaky material can be tolerated by adjustment of the proportion of fine to coarse aggregate to reduce void content, but when excessive or variable may produce difficulties in pumping, placing, compacting and finishing.

To minimize placing problems, it is important for aggregate producers to control the crushing process to minimize variation in shape between consignments of aggregates.

1.1.5 *Water absorption*

The water absorption of an aggregate is related to its particle porosity. Typical values of water absorption for natural aggregate lie in the range 0.5–5% by mass. The porosity can be calculated as relative density (OD) × absorption percent. For example, if the relative density (OD) is 2.50 and the water absorption is 3%, the porosity is $2.50 \times 3 = 7.5\%$ by volume.

Water absorbed into aggregate particles contributes to their mass but takes no part in providing concrete with workability and is not part of the voidage of the cement paste. Thus, absorbed water does not reduce strength by its presence. Indeed it may be beneficial as a reservoir of moisture to assist later hydration of cement. A limit on aggregate absorption is difficult to justify and is only rarely specified.

1.1.6 *Relative density*

Relative density (specific gravity) of an aggregate can be best understood as the ratio of the mass per cubic metre of the aggregate (W) to the mass of the same volume of water (V) (1000 kg/m^3). There are three relative densities of aggregate, illustrated in Table 1.8, which may be used in concrete calculations. Of these, relative density on a Saturated and Surface Dried basis is possibly the most relevant but its measurement is less precise than Apparent Relative Density. Apparent relative density is always the highest of the three, and relative density on an oven-dried basis is always the lowest. If the water

Table 1.8 Relative densities of aggregates. W, mass per m^3 of aggregate; V, mass of same volume of water; abs, water absorption

Apparent Relative Density	ARD		Dry weight ⎯⎯⎯ Dry volume	$\dfrac{W}{V} \times \dfrac{1}{1000}$
Relative Density on a Saturated and Surface-Dried Basis	RD$_{SSD}$		Wet weight ⎯⎯⎯ Wet volume	$\dfrac{W + \text{abs}}{V + \text{abs}} \times \dfrac{1}{1000}$
Relative Density on an Oven-Dried Basis	RD$_{OD}$		Dry weight ⎯⎯⎯ Wet volume	$\dfrac{W}{V + \text{abs}} \times \dfrac{1}{1000}$

absorption is large then the difference between each of the three is also relatively large. The three values are related to absorption and to each other by the following formulae, which can be used to check that results of tests are mutually consistent with one another:

$$\text{abs}(\%) = 100 \times \left[\frac{1}{RD_{OD}} - \frac{1}{ARD} \right]$$

$$\text{abs}(\%) = 100 \times \left[\frac{RD_{SSD}}{RD_{OD}} - 1 \right]$$

$$\text{abs}(\%) = 100 \times \left[\frac{1 - \dfrac{RD_{SSD}}{ARD}}{RD_{SSD} - 1} \right]$$

Example
abs = 2% by mass
ARD = 2.66
RD_{SSD} = 2.58
RD_{OD} = 2.53

Are these values reasonably consistent with one another?

Answer
Yes, they are. All formulae above give similar values for absorption, 2% approx.

Typical values for relative density of common dense aggregates lie in the range 2.4–2.8. Relative density, particularly that on an SSD basis, is used to convert aggregate mass to volume when calculating the theoretical volume of concrete produced from a given mass of materials. Relative density is not normally specified. However, knowledge of relative density is helpful in assessing the potential for a high-density aggregate to meet a specified density of the concrete for special purposes, e.g. shielding from radioactivity.

1.1.7 Bulk density, void content and voids ratio

Bulk density of dry or damp, compacted or uncompacted aggregate has relevance in some approaches to concrete mix design [12, 16], for conversion from bulk volume to weight, for purchase and stock calculations and for weigh-hopper and storage-container design.

Bulk density is less for uncompacted than for compacted aggregate and for damp compared with dry or saturated aggregate. This latter effect, illustrated by Fig. 1.3, is associated with a phenomenon called bulking, which has its maximum effect at low moisture contents when aggregate particles will be held apart by surface tension of the water film. Bulking is of relevance in design of containers for storing damp materials.

Because of the variation of bulk density with moisture and compaction, weigh batching of aggregates is preferred to volume batching. Indeed, volume batching is an anachronism today, except for some lightweight aggregates

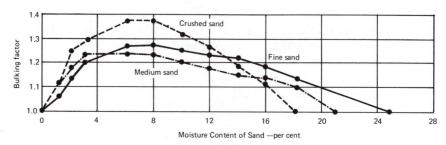

Figure 1.3 Bulking factor for sands with different moisture contents. After Neville [130]

having low particle densities and high water absorptions, which may be batched more accurately by bulk volume.

For dry normal-weight aggregates, bulk density can vary from 1000–2000 kg/m³ dependent on grading, shape, surface texture and degree of effort in compaction. An uncrushed gravel will normally have a higher loose bulk density than a crushed rock aggregate of the same grading, but the difference will be smaller for the compacted than the uncompacted state.

Bulk densities of different aggregates cannot be compared usefully without taking account of the densities of the materials of which each is composed. Void content measurement is a way of overcoming this problem. The void content can be assessed from the bulk density and relative density as follows:

$$\text{void content } (\%) = 100\left[1 - \frac{(\text{bulk density on an SSD basis})}{(1000 \times \text{relative density on an SSD basis})}\right]$$

or

$$100\left[1 - \frac{(\text{bulk density on an OD basis})}{(1000 \times \text{relative density on an OD basis})}\right]$$

For example, if the OD bulk density of a coarse aggregate is 1500 kg/m³ and the OD relative density is 2.50, then the voids content is

$$100\left[1 - \frac{1500}{1000 \times 2.50}\right]\%$$

$$= 100(1 - 0.60) = 40\%.$$

Void content is the percentage of voids, discounting particle pores, in the total volume of aggregate. Values for different aggregates of a similar grading can be compared directly. For the same grading, materials of lower void content can be expected to require a lower water content in concrete. When the gradings differ, comparisons cannot be made directly, and complex analysis (or, more usually, concrete trial mixes) is necessary to compare materials.

At low compactive effort, lower void contents are associated with a wide range of particle size and with round smooth particles. This implies that aggregate of large maximum size and rounded natural gravels and sands tend to have lower water demands in concrete than crushed rocks, crushed rock fines, flaky aggregates and small maximum size aggregates of any type.

In one approach to mix design [12, 16], the term 'voids ratio' is used; this is defined as the ratio of voids to the volume of solid material and can be calculated from bulk density and relative density as follows:

$$\text{voids ratio} = \frac{1000 \times RD_{OD \text{ or } SSD}}{BD_{OD \text{ or } SSD}} - 1$$

For example, using the same data as in the previous example,

$$\text{voids ratio} = \frac{1000 \times 2.5}{1500} - 1 = 1.67 - 1 = 0.67$$

Void content and voids ratio are related to each other as follows:

$$\text{voids ratio} = \frac{\text{voids content}}{1 - \text{voids content}}.$$

1.1.8 Moisture content

Moisture content is a function of the natural state and the processing of aggregates. Gravels and sands, whether dredged or excavated dry, usually need to be washed to remove unwanted excess silt and clay, and are usually stored for a time in the open, when they may collect rain water or drain, depending on weather conditions. Typical moisture contents for gravels and sands lie in the ranges 1–5% and 5–15% respectively. Crushed rock aggregates are normally processed dry, but may contain some moisture, particularly when stored in stockpiles.

Moisture content is of interest because of its contribution to the mass of damp aggregate and also to the water content of concrete. Knowledge of moisture content is important when purchasing aggregate by weight and when batching aggregate for concrete.

Moisture will drain through aggregates, and there will almost always be a moisture gradient through stockpiles and storage bins. Sampling of aggregates for moisture content needs to be thoroughly representative or related to specific bulk quantities if meaningful results are to be obtained.

Moisture content measurement can be based on free or total moisture, the difference being the absorbed water:

$$\begin{Bmatrix} \text{Free moisture} \\ \text{content (\%)} \\ \text{by mass} \end{Bmatrix} = \begin{Bmatrix} \text{Total moisture} \\ \text{content (\%)} \\ \text{by mass} \end{Bmatrix} - \begin{Bmatrix} \text{Aggregate} \\ \text{water absorption (\%)} \\ \text{by mass} \end{Bmatrix}$$

1.1.9 *Chloride content*

Chloride content of concrete may sometimes be restricted by specifications in respect of resistance to sulphate attack or steel corrosion, e.g. by BS 8110, as shown in Table 3.2 in section 3.4.5 on concrete durability.

BS 882 provides values, shown in Table 1.9, which may be adopted by the specifier who wishes to include chloride limits for the aggregate, but these are optional, not obligatory.

Chloride is a naturally-occurring ion in sea water and in some land aggregates. Marine-dredged aggregates can be expected to contain some chloride, but this is usually significantly reduced [7, 8], compared with sea water, by washing the aggregates with estuarine, river or mains water, as shown by Fig. 1.4, and may be further reduced by draining.

Other salts in sea water are not generally considered harmful to concrete [17]. However, the presence of sodium may need to be taken into account for

Table 1.9 Optional maximum chloride content limits provided in BS 882

Type or use of concrete	Maximum total chloride content expressed as percentage of chloride ion by mass of combined aggregates
Prestressed concrete and steam-cured structural concrete	0.02
Sulphate-resisting Portland cement concrete	0.04
Concrete containing embedded metal and made with cement complying with BS 12	0.06 for 95% of test results with no result greater than 0.08

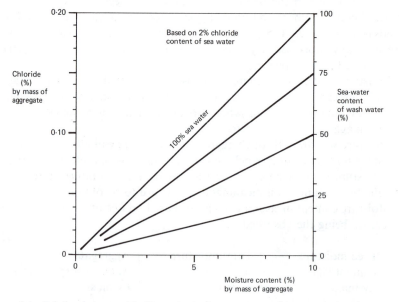

Figure 1.4 Relation between chloride content of aggregates, moisture content and sea-water content of wash water

alkali–silica reaction. For convenience, chloride measurements may be used to estimate reactive sodium oxide levels in aggregates.

1.1.10 *Deleterious materials*

There are some materials [18, 19] which may be considered deleterious in concreting aggregates, for example lignite and coal; however, small proportions of such materials can usually be tolerated. Indeed, where such inclusions occur it is very unlikely that an aggregate could be provided totally free of them because of the technical difficulties and cost of ensuring complete removal.

Tolerance is justifiable on the grounds that such residual material usually has only a slight weakening effect which is overcome by use of more cement in designed mixes. In some parts of the country, free mica for example may be present in quantity; this may provide delight to one architect who is attracted by the sparkle but concern to others who are aware of its platy shape. Again, competent mix design will ensure that any increase in water demand, loss of cohesion, and loss in strength are adequately compensated by a slight increase in cement and fine aggregate contents and there need be little concern over its presence [20, 21].

Obviously there will be cases where the specifier is particularly concerned over surface appearance or durability and may wish to use materials which are practically free from deleterious materials. It is important that the specifier makes clear his special needs at an early stage and understands that this may require importing of special materials from other areas, with consequent increased costs.

Possibly one of the most difficult materials to remove is iron pyrites, a natural material occurring in some parts of the country, which can decay on the surface of concrete to form unsightly brown rust stains.

Mine tailings need to be checked before use as a concrete aggregate. In the SW of England, the use of Killas (Devonian slate or shale) containing clay and Mundic (iron pyrites) has led to deleterious expansion of precast concrete blocks.

Some types of silica can react slowly with alkalis from cements to form an expansive gel which, under extreme conditions, might eventually disrupt concrete. As yet, an accepted BS test is not available for assessing the susceptibility of an aggregate to reaction with alkalis.

Certain rock types are recognized as being unlikely to react, either because they are silica-free or because any contained silica is unlikely to be in a reactive form [22, 145]. These are shown in Table 1.10. This list is, however, of limited value because the vast majority of sands contain silica in a range of forms. It will also be noted that two of the commonest coarse aggregates, flint (or chert) and quartzite, are not included in the list. Fortunately, it is now becoming accepted that where the total aggregate consists of 60% or more flint or chert, it can be treated as unreactive. Generally, the few cases of serious damage which have occurred have been restricted to the Midlands and the SW of England. Further information is provided in the cements section (Section 1.2.8).

Table 1.10 Aggregates categorized by the Concrete Society as unreactive with alkalis [22]

Air-cooled blastfurnace slag	Expanded clay/ shale/slate	Microgranite
		Quartz
Andesite	Feldspar	Schist
Basalt	Gabbro	Sintered pfa
	Gneiss	Slate
Diorite	Granite	Syenite
Dolorite	Limestone	Trachyte
Dolomite	Marble	Tuff

Note: Opaline silica must be absent

1.1.11 Shells

Shell is often considered erroneously as a deleterious material. Most shells are formed of calcium carbonate, the substance of limestone, having a similar relative density of about 2.70, which is higher than that of flint (2.65), and having a lower absorption. Broken and ground shell in sand is almost always smooth and polished, aiding workability. Its flat or slightly curved shape may be a slight disadvantage but is compensated for by its smoothness. Some land aggregates as well as marine aggregates contain shell, and such materials usually perform at least as well as aggregates without shell.

In the case of coarse aggregates, there may be some concern over hollow shells or large flaky shells [23]. In the case of hollow shells, cement paste usually enters and fills the voids [24]. In the case of large flaky shells, their presence is usually compensated for by a much more rounded shape of the remaining aggregate particles, so that no significant fears should remain concerning their use [7, 8, 24].

BS 882 includes limits on shell as follows: maximum 8% in the fraction coarser than 10 mm; maximum 20% in the 10 mm–5 mm fraction; no limit for fine aggregate. Although the limits are not technically necessary, they are not usually found to be too restrictive.

1.1.12 Uniformity

The most important features of aggregate for the ready-mixed concrete producer are uniformity within consignments and consistency over time, rather than the absolute value of any property, provided of course the values are in compliance with specifications [13, 25]. When an important property alters significantly, it is vital for the ready-mixed concrete producer either to be informed or to identify it for himself to minimize the risk of operating outside a specification requirement for the concrete.

The most important properties of aggregates affecting the properties of fresh concrete are:

Moisture content of fine aggregate
Grading of fine and coarse aggregate
Void contents of fine and coarse aggregate.

Moisture is critical because of its influence on the amount of water required to be added to concrete to produce the desired workability. Grading of fine aggregate is important because of its influence on the cohesiveness of concrete. Void contents of fine and coarse aggregate are important because of their influence on both cohesiveness and water demand of concrete. Void contents are influenced particularly by changes in grading and shape, both of which can be influenced, in the case of gravels and sands, by the extent to which oversize material is being crushed and the uniformity with which it is being blended with the uncrushed material.

1.1.13 *Non-standard aggregates*

It is difficult to write one standard for use of aggregates in concrete nationwide which provides safeguards for all potential users but at the same time does not place restrictions on experienced and knowledgeable concrete specifiers, contractors and ready-mixed concrete producers. To overcome this, BS 5328 and BS 8110 permit aggregates to be used which do not comply fully with aggregate standards, where there is agreement between the purchaser and supplier of concrete. This basis of such an agreement might be the evidence of satisfactory use in practice or the availability of appropriate test data, as permitted by BS 8110.

1.1.14 *Making the best use of natural resources*

The fewer unnecessary restrictions placed by specifiers upon aggregates, the more economic and effective use can be made of locally available materials.

1.1.15 *Moisture movement*

When concrete dries out or is wetted again, the cement paste attempts to shrink or expand. Usually this movement is adequately restrained by the skeletal aggregate structure. Indeed, this is an important function of the aggregate.

Most concreting aggregates have very low moisture movements, but certain types of aggregate, notably the more absorptive types of dolerite, have physical pore structures which are particularly susceptible to movement with changing moisture condition [26].

Moisture movement test methods have relatively poor reproducibility, and this should be taken into account when appraising test results. There is now a BS method, BS 812: Part 120, for testing and classifying drying shrinkage of aggregates in concrete.

Using this test method, drying shrinkages for flint, quartzite, limestone and granite concretes are typically in the range 0.01–0.04%, whereas values of over 0.07% can be obtained with some aggregates. BS 812: Part 120 and BRE Digest 357 [26] recommend that aggregates with drying shrinkage values above 0.075% should be limited in their use generally to situations where complete drying out never occurs or for structural members that are symmetrically and heavily reinforced and not exposed to the weather.

1.1.16 *Lightweight aggregates*

The commonest available types of lightweight aggregate for use in ready-mixed concrete are sintered pulverized fuel ash and expanded clay, but the following have also been used: expanded slate or shale, expanded or pelletized slag, natural pumice, exfoliated vermiculite, and expanded perlite.

Some materials are available as normal-sized coarse aggregates, but some may be available only in the smaller single sizes. Some coarse materials have rounded shape and smooth texture whereas others, particularly those crushed from clinker, are angular, rough and vesicular. Some materials are available as fine aggregate, usually obtained by crushing of coarse particles, so that the shape and texture of the fine aggregate are likely to be angular, rough and vesicular.

Particle densities can be very low, sometimes below unity, which means that particles can float on water, causing problems during mixing and transportation [27]. Segregation can be reduced by including an air-entraining agent in the concrete which can have the additional benefit of further lowering density [27].

Aggregate particles are often highly absorptive and may be readily friable, abradable or degradable. Water absorption values can approach or even exceed 100%, implying that particles can absorb more water than their own dry weight.

The range of possible densities and applications of lightweight aggregate concretes [28, 29] (Table 1.11) can be widened by suitable selection from the following techniques, separately or in combination:

No fines
Low workability, partial compaction
Normal workability, full compaction
High or low cement content
Lightweight or normal-density fine aggregate
Air entrainment.

In producing ready-mixed lightweight aggregate concrete, the following need to be considered.

Table 1.11 A range of lightweight concrete densities and strengths [28, 29]

Types of lightweight aggregates	Air-dry concrete density (kg/m^3)	28-day cube strength (N/mm^2)
Sintered pulverized-fuel ash	1500–1900	15–55
Foamed slag	1400–2000	7–45
Pelletized expanded slag	1400–1900	7–45
Expanded shale	1400–1800	20–45
Expanded clay	1100–1600	7–15
Perlite	400–650	1–5
Vermiculite	400–550	0.5–1

  (i) Very light and dry particles can blow off stockpiles and belt conveyors
 (ii) Stockpiles and bin contents may need heavy hosing and draining to ensure stable moisture content
(iii) Additional ground storage or overhead bins may be needed. Movement of aggregate may require additional measures, e.g. normal vibration may not allow flow from hoppers
(iv) Accurate batching is difficult unless uniform and known moisture content is obtained—volume batching may be appropriate for very light materials
  (v) Mixing method may need modification—a longer mixing time may be necessary
(vi) Fast and excessive loss of workability can occur due to absorption by dry aggregates or to degrading of particles
(vii) Plastic and hardened density and yield are more variable than for dense aggregate concrete
(viii) Trucks cannot be loaded to full weight capacity
 (ix) Staff need time and training to become accustomed to unusual materials or mixes.

1.1.17 Heavyweight aggregates

The commonest available types of heavyweight coarse aggregates [30] are barytes (barium sulphate), ilmenite and magnetite (iron ores), and steel shot, although other varieties of iron ores and shots are sometimes used. High-density concrete is often required for shielding from radioactivity in medical or nuclear constructions. Using the above materials as coarse aggregates with natural sands, the densities shown in Table 1.12 might be achieved [30].

The following aspects need to be taken into account in producing high-density concrete.

  (i) Special attention is needed to control workability and to minimize segregation and bleeding with heavy coarse aggregate.

Table 1.12 Examples of concrete densities obtained with heavy coarse aggregate [30]

Coarse aggregate	Concrete density (kg/m^3)
Barytes	3000
Magnetite/ilmenite	3200
Iron shot	4700

Note: Higher densities are achievable if heavy fine aggregate is also used.

  (ii) Reduced volumetric capacity of storage bins, weighbatchers, conveyors, mixers and transporting vehicles for a given weight. Additional ground and bin storage may be needed.

 (iii) There is a possible need for higher-capacity balances for weighing test specimens, particularly for plastic density.

 (iv) Heavy coarse and fine aggregates may not be available with normal BS 882 gradings. In some instances, only coarse aggregate may be available.

  (v) Batch weights of materials will need adjusting to take account of higher density of coarse aggregate and the capacity of the equipment.

 (vi) Some aggregates may degrade during mixing.

Water reducing agents may assist segregation problems and marginally increase density.

1.2 Cementitious materials

Cementitious materials, or hydraulic binders, are obviously key components of concrete. In fresh concrete, they are the finest particles and they have the function of providing workability, cohesion and stability. In the hardened concrete they have responsibility for providing mechanical strength, a relatively impermeable pore structure and an internal environment to resist chemical attack of the concrete and to protect any steel reinforcement.

In order to achieve these benefits, the various complex components of the cementitious materials react with water at various rates in concrete and form new products. This process, termed hydration, results in an interwoven physically bonded material structure, the components and porosity of which are dependent upon the types and activities of the cementitious materials, the amount of water (originally present), the internal moisture and temperature history and the age of the concrete. Some of the components react relatively quickly, while others take many years to complete their reactions, and indeed may never be permitted to achieve anywhere near their full potential because of the absence of a suitable stable internal environment.

The chemical reactions create heat, which in turn promotes increased rates of reaction, so that the process of stiffening and hardening is initially an accelerating one. To ensure that loss of workability is small during the first few hours, the main reactions are delayed by inclusion of a form of calcium sulphate in the Portland cement component. One benefit of the heat development and accelerating process is that a significant proportion of concrete's potential strength may be achieved in the heart of the concrete within a few days of construction. One disadvantage of the otherwise slow process of hydration is that unless the correct internal environment is maintained throughout, for a long time—many weeks or months—the key surface zone for durability may achieve only a fraction of its potential strength and impermeability.

The properties of cementitious materials which have the greatest relevance for the ready-mixed concrete producer are:

Grading, mean size and fineness
Relative density
Water demand
Setting times
Strength in concrete at 28 days
Ratio of 7- to 28-day concrete strength.

Other properties which have particular relevance for meeting certain specification requirements are sulphate resistance, alkali content, chloride content and colour.

The commonest cementitious materials in use in ready-mixed concrete are:

Portland cement (pc)
Sulphate-resisting Portland cement (srpc)
Portland blastfurnace cement
Portland pulverized-fuel ash cement

Ground granulated blastfurnace slag (ggbs) ⎫ in combination with
Pulverized-fuel ash (pfa)   ⎬ Portland cement.
                                            ⎭

1.2.1 Grading, mean size and fineness

A typical grading for Portland cement might be as shown in Table 1.13. The mean size is about 0.015 mm (15 microns), compared with about 0.5 mm for a sand and 10 mm for a graded 20 mm gravel, so that there is an appreciable size differential between the three major solid components of concrete. This size differential minimizes particle interference and enables the sand to fill the voids in the gravel and the cement to fill the voids in the sand, leaving just the voids in the cement to be filled by water and air.

The fineness of cement is measured as a surface area per unit mass which bears some relation to mean size, as well as to grading, particle shape and

Table 1.13 Typical grading of cement [31]

Size (micron)	% passing
300	100
125	99
63	93
32	77
16	54
8	37
4	21
2	10

texture. A typical range of values for fineness of Portland cement is 290–390 m^2/kg [32].

For ready-mixed concrete, the cement grading or fineness has a significant influence on water demand, cohesiveness, bleeding, surface finish and rate of gain of strength. Finer cements tend to have higher water demands in richer mixes, but have increased cohesiveness, reduced bleeding, enable a more uniform surface finish to be achieved and produce higher earlier strengths. In fact, a rapid-hardening Portland cement is usually just a more finely-ground Portland cement. Ggbs and pfa have gradings of a generally similar order to those of Portland cement.

1.2.2 Relative density

Typical relative densities for cementitious materials (pc, sulphate-resisting cement, ground granulated blastfurnace slag and pulverized-fuel ash) are shown in Table 1.14.

1.2.3 Water demand

The water demand of concrete [12, 16] can be related to the cement particle size distribution and to the water demand of cement paste. This latter property is measured by a test for standard consistency using the Vicat apparatus. This test involves determining the water content producing a certain plunger penetration in a small sample of freshly compacted paste.

In effect, the standard consistence test is a bulk density test like that for aggregates, except the medium is not air but water, which is obviously

Table 1.14 Typical relative densities for cementitious materials

pc, srpc	3.12
ggbs	2.8
pfa	2.2–2.4

appropriate because that is the medium in which cement is required to operate in concrete.

A typical range of values for water content in the Vicat test is 24–28% by mass, i.e. water/cement of 0.24–0.28. This means a voids content of about 0.5 (or a voids ratio near unity) which, considering the wide gradings of cement, is rather high and suggests rather angular or irregular-shaped particles, which is confirmed by photomicrographs. The voidage is also higher due to some agglomeration of cement particles.

In the case of ggbs, similar or slightly lower water demands and void contents are usually obtained, while for pfa, appreciably lower values occur, associated with the round, smooth shape of pfa particles. These help to explain why concretes containing pfa (and to a lesser extent ggbs) can show markedly lower water demands and also greater cohesion.

When using cements of a given type having high water demands, higher cement contents will be required in designed mixes, for a given workability and strength.

1.2.4 *Setting times*

The historic terms set or setting are very misleading, suggesting that there exists a time, or times, when cement paste or concrete suddenly develops different characteristics. In fact, the so-called initial and final sets are merely two arbitrary points in time on the continuous curve of stiffening at which particular needles, used in the Vicat test for cement paste, penetrate the paste to particular extents. This is shown diagrammatically in Fig. 1.5.

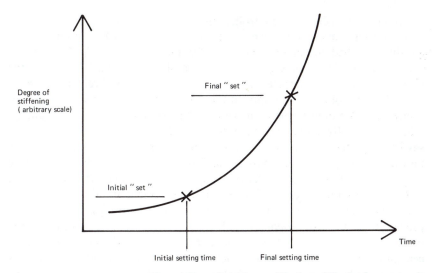

Figure 1.5   Illustration of the arbitrary nature of initial set and final set of Portland cement paste

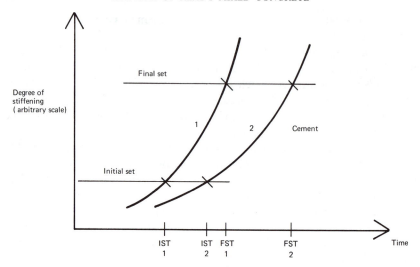

Figure 1.6   The valid use of initial and final setting times to compare rates of stiffening of different cements

Setting times serve a useful purpose in enabling the performance of two cements to be compared, as in Fig. 1.6. Thus, by comparison of setting times, cement 1 is seen to be a faster-stiffening cement than cement 2, and this may have significance for concrete, particularly with the richer mixes.

It does not necessarily follow that fast-stiffening cements have higher earlier strengths, although the assumption can sometimes be valid. Similarly, it does not necessarily follow that rapid-hardening Portland cements stiffen more rapidly.

Cements are generally controlled in their manufacture in such a way that setting times for cement pastes are typically about $1\frac{1}{2}$–4 h for the so called initial set and 3–5 h for the final set [32]. These times cannot be translated directly to concrete, as shown in Fig. 1.7, because the rates of stiffening of concrete will depend on the water and cement contents. Also, the rates of heat development and hydration are much greater in neat cement pastes. For example, a mix with a cement content of 200 kg/m^3 and an initial slump of 100 mm might still be workable 5 h after mixing compared with say 2 h for cement paste. Environmental conditions, particularly temperature, may significantly affect the comparison. Cement setting times should therefore be used with caution, and preferably only for comparison purposes between cements. Only initial setting time is limited in cement standards.

It is an interesting point to contemplate that if hydration is a very long process, then at the microscopic or submicroscopic level there are always parts of concrete which are stiffening and hardening. Indeed, some unhydrated cement particles are almost always detectable, even in very old concrete. This is, at one and the same time, both heartening and discouraging; heartening

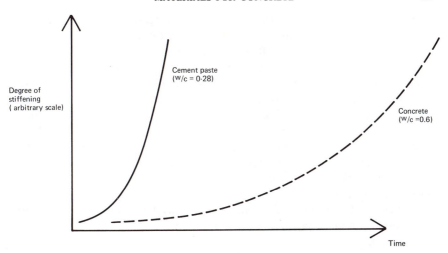

Figure 1.7 Illustration of different rates of stiffening for cement paste and concrete made with the same cement

because concrete may always have the capacity for further gain in strength; but discouraging in that it is not possible to fully utilize what has been purchased.

Despite the insistence upon the smooth curve of stiffening, there are possibilities for fast or almost instantaneous stiffening. One of these is 'flash' set, which can be associated with faults in the cement-making process, but is difficult to distinguish from other influences. Fortunately, it is rare for UK cements to exhibit this problem. 'False' set is slightly more likely, but still rare, the commonest cause being too high a grinding temperature. If a sample has been retained, it can be tested for false set using the American test in ASTM-C 451 [33]. Both false and flash set are considered again under section 2.2 on workability of concrete. When false set occurs, there is a fast but temporary loss of workability which can be regained if the concrete is remixed or vibrated. The inclusion of ggbs or pfa will usually increase stiffening times.

1.2.5 *Strength at 28 days*

The strength of cement is obviously an important property for ready-mixed concrete.

The standard test relates to a mortar prism having a W/C of 0.50. Very approximately, concrete of 0.60 W/C will have a strength of 0.8 × the standard mortar strength. Thus, 40 N/mm² concrete cube strength corresponds to a standard mortar prism strength of about 50 N/mm² at the lower W/C.

Cements are classed in terms of characteristic mortar prism strengths, the commonest standard values being 32.5, 42.5 and 52.5 N/mm².

The mean strength of a part of the cement consignment used in a batch of ready-mixed concrete, typically about 1.5 or 2 tonnes, can be expected to deviate from the short-term bulk average strength by about $\pm\,3\text{--}5\,\text{N/mm}^2$, so that it is not surprising that the variation in the strength of cement is obviously a major factor influencing the uniformity of strength of ready-mixed concrete.

The reporting of bulk average values, over a period of time, provides some extremely valuable information. It enables the ready-mixed concrete producer to confirm observed changes in mean level of concrete strength in his own control system. Even more important, due to the excellent co-operation between the cement and ready-mixed concrete industries, it has been possible for cement works to predict changes and to give advance warning to ready-mixed concrete producers.

The types of variation which can be experienced are illustrated in Fig. 1.8 a–d, and need to be accounted for in concrete producers' control systems and mix design.

The inclusion of ggbs or pfa may modify the 28-day strength of concrete

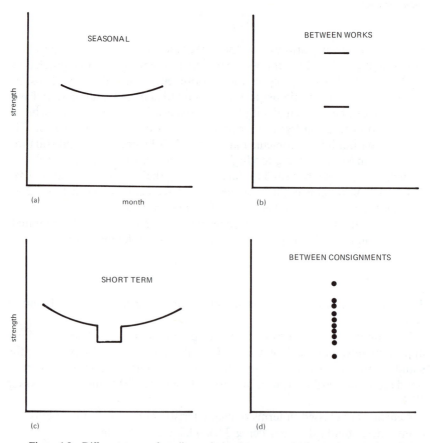

Figure 1.8 Different types of quality variation in cement which may be experienced

due to differences in cementing behaviour and due to influence on water demand. Effects may vary with proportions, type and sources of materials. However, the inclusion of ggbs or pfa does not usually result in increased variability of concrete. Indeed, reduced variability is often claimed.

1.2.6 Ratio of early to 28-day strength

Because of the 28-day delay between making concrete and obtaining a strength test result at the specified age, it is normal for the ready-mixed concrete producer to make an early age test, usually at 7 days, or an accelerated test at 1 to 3 days, from which the 28-day strength is predicted.

Typical UK average strengths for pc to BS 12 in concrete at a water/cement ratio of 0.60 are as follows:

3d   25 N/mm^2
7d   34 N/mm^2
28d   44 N/mm^2.

The relationship between 7- and 28-day strength of concrete is complex, but can be characterized through knowledge of the ratio of 7- to 28-day strength in the standard cement test of concrete. The ratio may vary from 0.5 to 0.85, dependent on cement type, source and production conditions, and thus can be significantly different from the value of 0.67 (2/3) commonly assumed by specifiers.

When the ratio is relatively constant, then accurate predictions can be made. If it changes, then predictions will become inaccurate until the change has been notified or detected. Naturally, it will be at least 28 days before the first ratio indicating a change can be seen by the ready-mixed concrete producer, so that prediction of a change and its notification by the cement works is vital. Again, this is being developed within the system of exchange of information between the cement industry and the ready-mixed concrete industry. An appreciation of this variation in 7- to 28-day strength ratio is also important for the concrete user, who may also be predicting 28-day from 7-day strength.

Composite cements or combinations of pc and ggbs or pfa can be expected to have lower ratios than for pc alone.

1.2.7 Sulphate resistance

BS 5328 provides advice on the use of pc, srpc, composite cements, ggbs and pfa, as summarized in Table 1.15. BRE Digest 363 provides alternative advice and indicates benefits at higher sulphate concentrations of using composite cements and combinations of pc with ggbs or pfa [44].

1.2.8 Alkali content

The sodium and potassium oxide contents of cement are expressed as a single

Table 1.15 BS 5328 recommendations* for use of various cements or combinations of cements with ggbs or pfa with 20 mm aggregates in concrete subject to sulphate attack

Cements and combinations of cementitious materials		Portland cement pc	Portland blastfurnace cement or pc and ggbs	High slag blastfurnace cement or pc and ggbs		Portland pfa cement or pc and pfa		Sulphate-resisting Portland cement srpc
BS		12	146	4246		6588	6588/6610	4027
Proportions of ggbs or pfa as percent of total cementitious materials		0	0–65	50–69	70–90	15–24	25–40	0
Min. cementitious material content (kg/m³) for class of Sulphate condition 1–5	1 Plain	275	275	275	275	275	275	(275)
	1 Reinf.	300	300	300	300	300	300	(300)
	2	330	330	330	310	330	310	280
	3				380		380	330
	4							370
	5							370

*See BRE Digest 363 [44] for alternative recommendations

value for equivalent sodium oxide:

$$\% \, Na_2O + (0.658 \times \% \, K_2O)$$

and known as the alkali content of cement. This is important because of its influence on the potential for alkali–silica reaction (ASR) between cement and aggregate in concrete. There is no specified BS limit for alkali content of cements, although maximum values may be declared by cement-makers.

Cements are considered to be unlikely to react harmfully [22] if the alkali content is less than 0.6%, provided active alkalis from other sources do not exceed 0.2%. The value of 0.6% cannot be guaranteed for Portland cements, and applies only for certain sources of sulphate-resisting Portland cements. This guarantee needs to be requested specifically at the time of ordering the cement.

An alternative view [22] is that, irrespective of the alkali level in the cement, alkalis are not significantly harmful, provided the alkali content of the concrete is below 3 kg/m³, with the benefit that cements of high alkali content can be accepted if the cement content does not exceed a corresponding value, taking account of variation above the monthly mean value for alkali content declared by the cement makers.

Table 1.16 provides maximum cement contents corresponding to the range of alkali levels observed in the UK. The average level in the cement was reported to be 0.65% in 1985, ranging from 0.4 to 0.9% [34].

The alkali contents of ggbs and pfa are assumed to contribute only partially to alkali–silica reaction. Ggbs or pfa in excess of 25% by mass of cement can be assumed to act as an alkali diluent and is often recommended as a means of reducing the risks of ASR. Ggbs may also act positively in some special way to inhibit the reaction, such that the inclusion of 50% or more ggbs is accepted as a means of avoiding ASR, irrespective of the alkali content of the cement or the

Table 1.16 Maximum Portland cement content in concrete at 3 kg/m³ maximum alkali level [22]

Declared monthly mean Portland cement alkali content (%)	Maximum target mean Portland cement content (kg/m³)
0.5	600
0.6	500
0.7	430
0.8	375
0.9	335
1.0	300

Note: Values include allowances for variation in monthly mean alkali content of cement and variation in cement content of concrete

concrete [22]. Further guidance is given in section 3.4.6, dealing with alkali–silica reaction.

1.2.9 *Chloride content*

Cement may contain up to 0.1% chloride as a minor constituent which needs to be taken into account when a maximum chloride content of concrete is specified.

1.2.10 *Colour*

The density or shade of the grey colour of cementitious materials is the most important factor determining the colour of the concrete. There is no standard requirement for colour or shade of cementitious materials.

Changes in colour of concrete associated with the cementitious materials may be observed when:

(i) Sources of cementitious materials are changed
(ii) Variations occur in the cementitious materials
(iii) Variations occur in the proportions of cementitious components of composite cements or combinations blended in the concrete mixer.

Generally srpc is a darker grey than pc from the same works, but some pc cements can be quite dark. Concretes with pfa may be darker, whereas those with ggbs are often lighter. Ggbs may produce a characteristic 'blue' shade for a short period early in the life of the concrete, compared with a greenish tinge for pc concrete.

1.2.11 *Ground granulated blastfurnace slag and pulverized-fuel ash*

Reference has been made to certain aspects of ggbs and pfa under the various properties of cements to enable direct comparison, but in view of the accepted and growing use of those materials, it is appropriate for them to receive a fuller consideration separately from Portland cements.

It is important to recognize the significant role they can and do have in increasing the range of composites. In many instances they enhance the concrete properties, and do more than act as replacement materials–a belittling and anachronistic view of their role.

Figure 1.9 shows in an idealized form how the use of ggbs in combination with Portland cement to BS 12 can result in an enhanced concrete property because of favourable chemical or physical interactions.

It is equally important to recognize that different sources of cement and ggbs or pfa will interact differently, such that each combination needs to be assessed separately and monitored continuously. For these reasons care is taken generally in this section to avoid suggesting specific quantitative

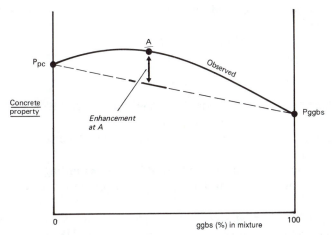

Figure 1.9 Idealized representation of the enhancement of properties of ggbs–Portland cement composites associated with beneficial interactions between the composites. Adapted from Osbaeck [110]

benefits. Where numbers are quoted, they are usually those resulting from consensus and incorporated into standards and codes.

The use of ggbs and pfa in concrete has increased in the UK in recent years and currently continues to do so for a number of significant reasons [35–38], including:

(i) Developments by suppliers of ggbs and pfa
(ii) Favourable relative price of ggbs and pfa compared with cement
(iii) Incentives for energy saving on a national basis
(iv) Research and development by the Building Research Establishment, CEGB, the Steel Industry and individual firms
(v) Existence of British Standards for cement incorporating ggbs
  (Portland blastfurnace cement, BS 146;
  high slag blastfurnace cement, BS 4246)
(vi) Production of Portland blastfurnace cements in Scotland, Northern Ireland and SW England
(vii) Production of Portland pfa cements in UK
(viii) Existence of a British standard for pfa, BS 3892.
(ix) Benefits of temperature reduction in massive pours; sulphate resistance, resistance to alkali–silica reaction
(x) Cement shortage in some parts of the UK in earlier years
(xi) Agrément Board Certificates for selected ggbs and pfa
(xii) Quality assurance schemes developed by the ready-mixed concrete industry in conjunction with manufacturers of the ggbs and pfa

(xiii) Experience of the ready-mixed concrete industry of ggbs and pfa concretes

(xiv) Use by major specifiers and contractors on prestigious contracts

 (xv) Use of ggbs and pfa elsewhere in the world.

Their acceptability has been increased further by

  (i) Publication of British Standards BS 6699 for ggbs and BS 6588 and 6610 for Portland pfa and pozzolanic cements

 (ii) Publication of BS 8110 which makes significant reference to the acceptability of ggbs and pfa

(iii) Publication of an amendment to BS 5328 permitting ggbs and pfa combinations with opc to have parity with composite cements [39].

1.2.11.1 *Specific advantages of ggbs and pfa concretes.* Composite cement, ggbs or pfa concretes are generally accepted in the UK to have specific advantages as follows:

  (i) *Sulphate resistance*
     BRE Digest 363 [44] advises that blends of ggbs or pfa with Portland cement can be used in a range of sulphate conditions. An increased cementitious materials content is required compared with srpc concrete. In some cases blends are permitted where pc by itself is not allowed. (See Table 1.15).

 (ii) *Alkali–silica reaction*
     Use of 25% or more ggbs or pfa coupled with a restriction on the alkali content of concrete, or the use of 50% or more ggbs with a limit of 1.1% on the alkali content of the combination of cement and ggbs, is accepted as providing a high degree of immunity against alkali–silica reaction such that the potential activity of the aggregate can be ignored [22, 145].

(iii) *Temperature rise in massive construction*
     Considerable benefit has been obtained in major UK construction by significantly reducing the early temperature rise of mass concrete by using ggbs or pfa while maintaining satisfactory strength development [40] and at considerable cost savings, both directly and also indirectly through faster construction.

1.2.11.2 *Ggbs or pfa in cement or added in concrete? A choice.* The limited availability of Portland blastfurnace cement or Portland pfa cement, and the need for versatility in meeting a wide range of needs and specifications, has meant that the ready-mixed concrete industry has been more inclined to use ggbs or pfa added directly to the concrete at the mixing stage rather than already pre-blended in the cement.

The obvious benefits of stocking a Portland blastfurnace cement or a

Table 1.17 Strength requirements for common classes of cements to BS 12, 146, 4246, 6588 and 6610 and for blends of Portland Cement to BS 12 and ggbs or pfa

	Portland	Portland blastfurnace	High slag blastfurnace	Portland pfa cement	Pozzolanic cement
Standard	BS 12	BS 146	BS 4246	BS 6588	BS 6588/6610
% ggbs or pfa	0–5	6–62 ggbs	47–81 ggbs	6–27 pfa	max 50 pfa
Age (days)			Strength(N/mm^2)		
2	10	10	—	—	—
7	—	—	12	16	12
28	42.5	42.5	32.5	32.5	22.5

Portland pfa cement are that the properties are guaranteed by the maker, who also undertakes the sampling and testing. Ready-mixed concrete, ggbs and pfa producers are often quite prepared to accept these responsibilities.

1.2.11.3 *British Standards and Codes of Practice relating to the use of ggbs, pfa and composite cements*. Table 1.17 provides a comparison of the specification requirements for various cements and corresponding blends in the concrete mixer.

BS 5328 for concrete permits ggbs and pfa to BS 3892: Part 1 to be treated as part of the cement content when the combination meets the requirements for proportions and properties of the appropriate British standard for a composite cement [39]. For many years Codes of Practice, e.g. CP 110 and CP 114, have accepted Portland blastfurnace cement as being on a par with Portland cement.

BS 8110 confirms the position of combinations as being on a par with the corresponding dry-blended or interground composite cement. BS 8110 also accepts cements and combinations to BS 12, 146 and 6588 as having equality with each other for normal durability situations, providing the same strength grades of concrete are used.

(i) *BS 6699 for ground granulated blastfurnace slag*
   The requirements of BS 6699 include chemical composition, glass count and fineness, all of which are subject to control by the manufacturer. Of these, fineness is probably the most important, to ensure reactivity and water reduction.
(ii) *BS 3892 for pulverized-fuel ash*
   BS 3892 has been revised to take into account the experience gained since its initial drafting in 1965, including the use of grading on the 45-micron sieve and the requirement of the concrete industry for a standard

providing greater control when pfa is to be used as a cementitious material in structural concrete. Part 1 of BS 3892 covers this use and includes a water demand requirement and also an optional test for pozzolanic activity. Eventually it is possible that the latter test may become obligatory, but there is insufficient experience to allow it to be so, at the present time.

The reasons for the interest in water demand and pozzolanic activity are to ensure that (i) the pfa is behaving as a cementitious material and not merely as a filler, and that (ii) lack of early cementitious action is compensated, at least partially, by reduction in water demand. BS 3892: Part 2, for general uses of pfa in concrete, adopts the main features of Part 1, but relaxes the loss on ignition and grading requirements. This material is intended for such uses as unreinforced precast concrete products, lean concrete and for grouts and mortars. It is not permitted by BS 8110 or BS 5328 to be treated as part of the cement content for durability (see also 1.4.1).

1.2.11.4 *Quality scheme for ready-mixed concrete–requirements for ggbs and pfa.* It is possible, by the rules of the QSRMC [41], to use ggbs or pfa either as a cementitious material or solely as an addition with no assumed contribution as a cementitious material. In the former case, equality of the combination against a cement standard is required, whereas in the latter case the ggbs or pfa is assumed to be behaving as an inert material only. Because of the wide range of proportions available through the use of the two blastfurnace cement standards, it is rare for ggbs to be used other than as a cementitious material. At present, pfa to BS 3892: Part 1 covers the use as a cementitious material; BS 3892: Part 2 material may not be used as a cementitious material, but is permitted as an addition.

Many clauses are included to provide assurances to clients that ggbs and pfa are used only under rigid control and only where permitted by the client. Crucial to the whole system are the requirements for sampling and testing for compliance against cement standards in respect of proportions and all properties.

Well before the latest developments included in BS 5328, the ready-mixed concrete industry was concerned to ensure that ggbs/cement combinations were only used when checks were made to confirm compliance against BS 146 or BS 4246. A scheme was developed for ggbs by which samples of cement were taken by ready-mixed concrete producers and sent to the ggbs producers for tests in combination with their materials. A similar scheme was more recently agreed with suppliers of pfa to BS 3892: Part 1, and both schemes have now been consolidated into one [42, 43]. A summary of procedures [42, 43] for sampling and testing of combinations of Portland cement and ggbs or pfa against the compliance requirements for cements in BS 146, 4246, 6588 or 6610 is provided in Table 1.18.

A certificate is provided which summarizes the results of strength tests on

Table 1.18 BRMCA and BACMI requirements for tests of ggbs and pfa [42,43]

Initial sampling and testing to determine standard deviation
1. At least 40 spot samples are obtained, each from a different consignment, of the cement judged likely to be associated with the greatest variation in strength of a blend with ggbs or pfa.
2. At least 40 spot samples of ggbs or pfa are sampled over the same period.
3. 50/50 or 70/30 blends of the cement and ggbs or pfa samples are tested against the requirements of BS 146, 4246, 6588 or 6610.
4. For the results of the 40 or more tests for each parameter, the standard deviation is estimated.

Routine test for each combination of sources of cement and ggbs or pfa
1. Each month at least 8 spot samples are blended to obtain a bulk cement sample.
2. Each month at least 20 daily spot samples are blended to obtain a bulk ggbs or pfa sample.
3. 50/50 or 70/30 blends of the cement and ggbs or pfa are tested against the requirements of BS 146, 4246, 6588 or 6610 to determine the mean monthly values of each parameter.
4. If the mean value of any parameter is less than 3 × standard deviation from the particular BS limit, then the reduced maximum proportion of ggbs or pfa to cement is estimated which will ensure compliance with this requirement. The estimation is obtained by further testing or from pre-determined families of relationships.
5. Results are certificated at monthly intervals together with a statement of the maximum proportion of ggbs or pfa to ensure compliance when using the stated source of cement.

concrete and advises on the limitations of the proportions of ggbs or pfa to ensure compliance with BS 146, 4246, 6588 or 6610. So, for example, the concrete producer may be warned not to use more than, say, 50% ggbs in order to maintain parity with all BS 146 requirements, particularly early strength, although BS 146 permits up to 65% to be used. The method by which the maximum value is determined is illustrated in Fig. 1.10.

Appreciable safety margins are applied, based on measured uniformity of each property, in establishing the maximum permitted percentage of ggbs or pfa, as shown in Fig. 1.10. QSRMC assessors check that higher ggbs or pfa proportions are not being used by ready-mixed concrete producers.

Because 2- or 7-day strength is usually more critical than 28-day strength, or other parameters, in comparison with BS requirements, these control the percentage of ggbs or pfa in combinations. The 28-day results are very rarely, if ever, critical control parameters restricting the proportion of ggbs or pfa to cement which can be used in concrete.

1.2.11.5 *Handling of ggbs or pfa.* Some modifications are essential, and others may be necessary, for handling ggbs and pfa, more particularly for pfa, as follows:

(i) Separate silos for cement and ggbs or pfa
(ii) Means of ensuring that correct silo is used for storing ggbs or pfa
(iii) Typically only 20 tonnes of pfa to be stored in a 30-tonne cement silo
(iv) Separate venting of silos to prevent contamination and finer air filters are required for pfa silos

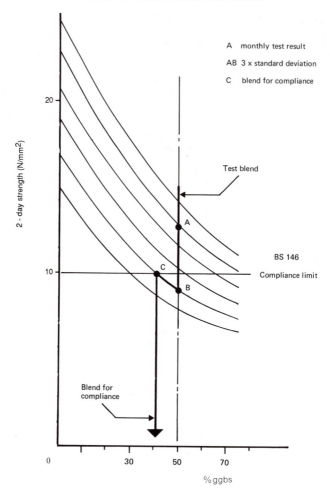

Figure 1.10 An example of the use of results from a monthly test blend to determine the maximum percentage ggbs in a combination with pc to BS 12

(v) Increased aeration and steeper outlets needed for pfa silos to prevent arching.

(vi) Replacement of air slide conveyors by screw feeders or bucket elevators for pfa

(vii) Ensuring that, with cumulative weighing, cement is always batched first and ggbs or pfa second

(viii) Training of batcherman in using new batch instructions and judging workability of ggbs and pfa concrete.

1.2.11.6 *Costs of using ggbs and pfa.* It is common misconception that any ggbs or pfa concrete is always cheaper than the same volume of plain concrete.

Certainly, some ggbs or pfa concretes may not only be cheaper than an equivalent plain concrete, but it may sometimes be the case that an equivalent plain concrete cannot be made with the available materials. In other cases the ggbs or pfa concretes may be dearer to produce than their equivalent plain concrete, and yet an overall benefit may accrue due to savings in construction or design costs, for example, savings on the need to cool or insulate concretes or enabling a thicker section to be used.

Misconceptions can arise when comparisons of price of equivalent mixes are made on the basis of materials costs alone, without taking into account the concrete producers' production and control costs, which naturally are not public knowledge.

These costs arise in some or all of the following areas:

 (i) Additional silo capacity
 (ii) Plant modifications to silos, transfer equipment, batching and weighing equipment
(iii) Additional batching operations
(iv) Laboratory trials and materials testing
 (v) Increased cementitious content relative to pc
(vi) Additional control testing.

Greater management control is needed to ensure correct usage of ggbs or pfa. Most batching plants need some modification, with associated increased costs, to store, transfer and batch ggbs or pfa, entailing particular consideration of silo capacity, filters, outlets, aeration, transfer systems and batching controls. Naturally, the influence of these costs on each cubic metre of concrete will depend upon the throughput of ggbs or pfa concrete and the regularity of its use.

1.2.11.7 *Influence of ggbs or pfa on water demand of concrete.* The inclusion of pfa will usually lead to a significant reduction in water demand of concrete; indeed, for pfa to qualify against BS 3892: Part 1, a water reduction is obligatory. In the case of ggbs, a small water reduction may also be obtained. Whether or not a reduction is obtained will depend on the characteristics of the cement, as well as on those of the ggbs or pfa, together with the effects of physical and chemical interactions between them.

1.2.11.8 *Use of ggbs and pfa to reduce mix temperature of concrete.* Ggbs and pfa have been used successfully [40] on many occasions to reduce the temperature build-up of concrete in thick sections and particularly when high strengths are required (see section 1.2.11.12 on strength under different curing conditions).

Such large pour construction requires collaboration from an early stage between specifier, contractor, ready-mixed concrete producer, cement maker and ggbs or pfa supplier. Collaboration is essential to ensure exchange of data, correct mix design, preplanning of pours and continuity of supply. The net result should be consistent qualities of the fresh concrete, low maximum temperature, absence of cold joints or cracking and attainment of the required strength of the hardened concrete in the structure.

1.2.11.9 Influence of ggbs or pfa on other properties of fresh concrete. It is often claimed, almost certainly correctly, that pfa concrete is more workable than a plain concrete of the same slump. Similar benefit may apply also to ggbs concretes. Pumping may also be effected more readily with pfa concrete, due to the greater cohesion.

The reduction in water demand reduces the tendency of pfa concretes to bleed, which reduces the risk of settlement and consequent settlement cracking. The reduction in bleeding implies a greater risk of plastic shrinkage stress due to drying, but this is resisted by a greater surface cohesion, so the risk may not be significant; in any case all types of concrete should be protected quickly from premature drying.

1.2.11.10 Use of admixtures in ggbs or pfa concretes. Admixtures, including air-entraining agents and superplasticizers, can be used in ggbs or pfa concretes, but the quantities of agent may need to be appreciably different from those used for plain concrete. This applies particularly to air entrainment, for which increased dosages are usually necessary when using pfa [45, 46]. See section 1.3 on admixtures and Table 1.22 for further information.

1.2.11.11 Special types of pfa concrete. Rolled lean concrete with high pfa content originally developed for dams in the UK and USA has been developed further into high pfa content rolled lean concrete for bases and high pfa content concrete of pavement quality for road surfacings [47]. By high pfa concrete is meant ratios as high as 50/50 or even 80/20 of pfa/cement. Such mixtures take advantage of the possibilities for optimum design with multiple components.

Some difficulties have been experienced in discharging concretes with high pfa concretes from the conventional ready-mixed concrete mixer drum.

1.2.11.12 Influence of ggbs and pfa on strength. Ggbs and pfa influence strength of concrete in at least two ways, firstly through influence on water demand, and secondly through hydration of the pfa and lime from the cement or hydration of ggbs activated by the cement. This reaction is slow to develop in the case of pfa; it may not be apparent for the first 7 or even 14 days, but becomes more significant by 28 days and later. In the case of ggbs, the hydration benefit is seen earlier.

The net result is that a ggbs or pfa concrete designed to have the same standard strength at 28 days can be expected to have a reduced relative strength before 28 days and a greater relative strength thereafter, compared with the plain pc concrete.

1.2.11.13 Influence of ggbs or pfa on strength under different curing conditions. The foregoing is based on standard curing, which is the normal basis for specification and purchase of ready-mixed concrete, but what is likely to occur in real structures?

There is a scarcity of information on some aspects of curing. For example, it might be expected that ggbs or pfa concretes would perform comparatively less well at later ages in slender sections allowed to dry before hydration had developed significantly. However, it appears, judging from available data for cube strengths under different curing conditions for concrete with and without ggbs or pfa, that problems are not significantly increased, provided the concrete is designed for the same strength grade. Figure 1.11 illustrates this point for 28-day strength for a range of pfa sources and qualities complying

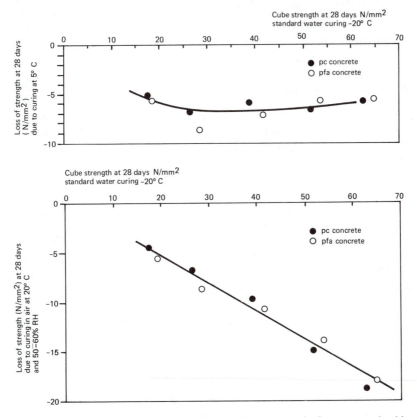

Figure 1.11 Influence of adverse curing conditions on 28-day strength of concrete made with pc or with pc and pfa to BS 3892: Part 1 or Part 2 grade A. Based on data from Dhir *et al.* [48]

with BS 3892: Part 1 and Part 2 Type A, when concrete is designed for the same slump and standard 28-day strength [48]. Despite this assurance, it would seem wise (such advice is often given) that ggbs or pfa concrete should be given even greater attention during early curing to ensure the quality at or near the surface. Indeed, BS 8110 makes this recommendation.

Ggbs and pfa concrete behave differently from pc concrete in thick sections likely to retain heat and allow build-up of high temperatures [40]. For plain concrete, the strengths achieved decrease as temperature increases, whereas ggbs and pfa concretes of similar standard strength increase in strength in the structure.

1.3 Admixtures

Admixtures are materials added to concrete, usually in relatively small proportions, to modify properties of concrete.

Admixtures have important roles [49, 50] in

(i) Enabling properties to be obtained which may be otherwise unobtainable
(ii) Enabling improvements in selected properties
(iii) Enabling more economic mix formulation
(iv) Overcoming problems in transporting, placing, compacting or finishing
(v) Overcoming deficiencies in other materials.

Their consideration increases the range of choices available to the specifier, contractor or ready-mixed concrete producer in seeking the most cost-effective way of meeting a specification or in seeking to overcome a problem.

It needs to be recognized that, in the UK, there is still some reluctance on the part of the specifier to permit admixtures to be used without express permission [50].

The reluctance is associated particularly with experiences, mainly with precast concrete, involving reinforcement corrosion and surface spalling caused by a combination of factors—inadequate cover, poor curing, or use of calcium chloride as an accelerator. Calcium chloride is not now permitted by BS 8110 to be used in concrete containing embedded metal. However, the introduction of BS 5075 (Concrete admixtures) is helping to build faith in the use of admixtures.

A second problem relates to the relatively small quantities of admixtures used. They are visible to the purchaser only in a few cases, e.g. pigments, air-entrainment). The purchaser may be concerned over uniformity of distribution through the batch and of the dangers of over-dosing. Such concerns are only overcome with satisfactory experience over a period of time.

BS 5328 and the QSRMC [41] permit admixtures to be used only with permission of the purchaser. Eventually, it is anticipated, as confidence is affirmed more specifiers will permit the ready-mixed concrete producer to take full responsibility without need to ask for permission.

The increasing availability of multipurpose admixtures may avoid the necessity to incorporate more than one admixture in the concrete, but may mean that optimization is not achieved, unless the formulation is specific to the combination of cement and aggregates with which it is to be used.

The range of commonly available admixtures [49] is indicated in Table 1.19. However, they are continually being added to, and potential users are advised to discuss problems with the Cement Admixtures Association or individual admixture manufacturers.

In addition to the main functions identified in their type descriptions, some admixtures can also be used to provide other benefits, as indicated in Table 1.20.

With the exception of solid pigments, admixtures do not normally contribute directly and significantly to the volume or weight of concrete and can be conveniently ignored in yield calculations. However, air-entraining and water-reducing agents and superplasticizers may have significant effects on either air content or water content or both, and any influences on yield need to be taken into account.

It may be necessary to take account of the chloride or alkali content of admixtures when maximum values are specified in respect of concrete or its components [22]. However, the contents are typically not likely to be significant except in the case of calcium chloride.

Table 1.19 British Standards for admixtures for use in concrete.

Type of admixtures	BS	
Accelerating	5075:	Part 1
Retarding	5075:	Part 1
Normal water-reducing	5075:	Part 1
Accelerated water-reducing	5075:	Part 1
Retarded water-reducing	5075:	Part 1
Air-entraining	5075:	Part 2
Superplasticizing	5075:	Part 3
Pigments	1014	

Table 1.20 Additional uses for admixtures.

Requirement	Admixtures
Reduce water content	Superplasticizing; air-entraining
Increase workability	Water-reducing; air-entraining
Increase early temperature	Accelerating
Reduce bleeding	Air-entraining
Reduce segregation	Air-entraining
Improve pumpability	Water-reducing; air-entraining; superplasticizing

Other aspects to consider include shelf-life of admixtures, special dispensers, necessity for remixing of stored admixtures to maintain uniformity, any special safety requirements, and sensitivity of properties to under- and over-dosing.

Most admixtures are sensitive to cement source and cement content, some to fine aggregate grading or content, and others, particularly air-entraining agents, may be affected by many other factors [45], so that formulation and dosage need careful attention to achieve the intended performance.

1.3.1 *Accelerators*

The prime benefit from using an accelerator is an increase in early strength. It will usually increase the rate of rise in early temperature and may also increase the rate of stiffening. This may be a disadvantage in hot weather but may be a benefit in winter, enabling earlier completion of finishing operations. While some additional frost protection may be obtained, it should not be assumed that concrete can be left unprotected in the winter or that long-term frost resistance will necessarily be improved.

The once common calcium chloride has now been largely replaced by non-chloride accelerators, which may be less effective at low temperatures and costlier, but do not present the risk of promoting steel corrosion in reinforced concrete.

In hot weather, use of an accelerator for high early strength may require mix adjustments to maintain slump at delivery because of increased loss of slump during transit.

1.3.2 *Retarders*

Retarders are in common use in concrete for slip-formed construction and in hot weather for extending the time available for placing. It should be noted that the rate of loss of workability associated solely with absorption or evaporation of water will not be reduced by the use of retarders. Because retarders will usually reduce early strength, it is common for compensating water-reducers to be incorporated. Dosage of retarders may need to be adjusted with ambient temperature.

1.3.3 *Water-reducers* (*normal and superplasticizers*)

Water-reducers have an important function in enabling specifications for particularly low maximum water–cement ratios to be met without exceeding specified maximum cement contents. They are also used extensively to enable higher workabilities to be used without violating max W/C requirements and without increasing bleeding or risks of segregation. Superplasticizers can be used as highly effective water-reducers. Rates of loss of workability may, however, be increased when water-reducers are used.

1.3.4 *Air entrainment*

The deliberate entrainment in concrete of air bubbles having a diameter of about 0.1 mm and ranging from 0.02 to 0.25 mm has a number of benefits.

(i)   To reduce water demand for particular workability
(ii) To increase cohesion and improve handling and finishing; pumping may be assisted
(iii) To reduce bleeding
(iv) To modify pore structure, giving lower permeability
(v)   To increase resistance to freeze–thaw cycles and to thermal shock associated with use of de-icing salts.

Although air-entrained concrete is more porous, the pore structure is discontinuous, provides a reservoir of pores for ice to form without disrupting the concrete, and reduces the flow of fluids into and through the pore system. Air-entrained concrete is thus likely to be less permeable and more durable under most conditions.

The increased porosity reduces strength but this is compensated for, partly or sometimes completely, by the reduction in water/cement ratio associated with reduced water demand. Generally, air-entrained lean concretes are stronger than plain concrete but the reverse usually applies to rich mixes requiring an increased cement content for the maintenance of strength.

However, any additional benefit of air-entraining concrete of high cement content is questionable. Control of air content is more difficult, avoidance of blowholes may be almost impossible, the possibility exists of obtaining a weakened layer near to the surface, and it may not be possible to comply with a strength grade requirement.

Some air-entraining agents are blended with water-reducing agents and are less prone to reduce strength, but may produce less stable air contents and increase the rate of loss of workability.

BS 5328 recommends different percentages of air for different maximum size of aggregates, as indicated in Table 1.21. It should be noted that the total air includes an allowance for entrapped air in fully compacted concrete. The test for air does not distinguish between entrapped and entrained air.

Table 1.21 Air contents recommended by BS 5328

Nominal max. aggregate size (mm)	Target total air content (%)
10	7.5
20	5.5
40	4.5

Use of air entrainment is not an alternative to good concreting practice; air-entrained concrete requires proper compaction and curing to achieve optimum performance. Finished surfaces should be able to drain to minimize the possibility of full saturation, because even air-entrained concrete is likely to be less frost-resistant when fully saturated. It is necessary also to take account of the increased yield in calculating batch weights, remembering also to account for the reduced water demand.

Air-entrainment can be particularly helpful in overcoming grading deficiencies in aggregates, the tiny bubbles acting as very fine particles, improving cohesion. With well-graded aggregates, fine aggregate content can usually be reduced when air is entrained.

When air-entrained concrete is used only rarely, difficulties in quality control can occur, but when produced regularly or in large quantities, close control can be maintained. The amount of air-entraining agent required to entrain a specific amount of air will vary with a number of factors [45], as shown in Table 1.22.

When an air-entraining agent is used, efficiency of mixing is especially important, because of the relatively small quantity used and because the air generated will depend upon mixing effort. The workability will be affected in turn by the amount of air generated. If plant mixing is inadequate, results of tests for air and slump at delivery may be significantly different from any made at the depot, because of the effects of agitation.

Air can be lost after delivery during subsequent transporting, placing and compacting, and it is therefore advisable to clarify in terms and conditions that

Table 1.22 Effects of various factors on air-entrainment. Adapted from Brown [45].

Factors producing a reduction in quantity of entrained air	Example of change	Typical effect on air content
Higher temperature*	$+10\,°C$	-1% air
Lower slump*	50 to 25 mm	-1% air
Higher cement content*	$+50\,kg/m^3$	$-\frac{1}{2}\%$ air
Coarser sand*	mid-zone F to M	$-\frac{1}{2}\%$ air
Lower sand content*	40 to 35% sand	$-\frac{1}{2}\%$ air
Organic impurities in sand	Present	Variable
Inclusion of pfa [46]	Present	Magnitude linked to carbon content and fineness of the pfa
Hard water*	Present	Reduction
Prolonged agitation of well-mixed concrete	$+1$ hour	$-\frac{1}{2}\%$ air

Note: increases in quantity of air will be associated with the opposite movement of factors marked*

the point of testing is at delivery. If the purchaser wishes to take into account any subsequent losses in air it is necessary for him to increase the specified percentage of air accordingly [51]. For example, pumping may lead to a reduction of air content or to modification of the bubble size distribution. Fortunately, the more important smaller-diameter bubbles are usually less destructible, and frost resistance may be maintained.

The use of pfa may require the dosage of air-entraining agent to be increased substantially to maintain the correct air content.

With experience, control of air content within $\pm 2\%$ of target value, as required by BS 5328, can be achieved, provided account is taken of the factors discussed. Close workability control is essential because of the interrelation between slump and air content.

Air-entrainment may be beneficial when used with lightweight aggregates as an aid to cohesion and in meeting a specified density.

BS 5328 identifies that air-entrained concrete with more than $350 \, \text{kg/m}^3$ of cement may be difficult to compact and finish, and that it may be difficult to achieve the required strength; BS 8110 advises trials.

1.3.5 Superplasticizers

Superplasticizers [52] are amongst the most exciting developments in concrete technology in recent years. They can be used to:

(i) Achieve substantial increase in workability without increase in water/cement ratio, segregation or loss of strength
(ii) Achieve substantial reduction in water/cement ratio
(iii) Produce flowing, self-compacting or self-levelling concrete.

The effectiveness of the admixture is usually limited to 30–45 minutes after addition of the admixture. When this applies, the admixture may need to be added to the batch at the construction site and the concrete remixed at normal mixing speed for at least 2 minutes. Alternatively, combined retarding superplasticizing agents might be adopted.

To achieve flowing concrete, 1.0–1.5% of admixture by mass of cement is added to concrete of 100–120 mm slump, the effect of the admixture being to increase the slump to 150–225 mm or a spread in the BS 1881 flow test of 550–620 mm. An increase in percentage fines is advisable compared with a normal mix, and a minimum total content of cement + fine aggregate (passing 300 microns) of $450 \, \text{kg/m}^3$ may be needed to minimize risk of segregation. After the initial period, appreciable water shedding or bleeding may occur, but homogeneity is normally maintained. Similar 28-day strengths to those of the unmodified mix can be expected.

Placing rates can be very high, and site handling needs to be well organized to take full advantage of the self-compacting and self-levelling qualities of flowing super-plasticized concrete.

For use of concrete in cast-in-place piling, the lower flow values in Table 2.4 may be considered more appropriate.

1.3.6 *Pigments*

Pigments are black or coloured fine powders, added dry or as suspensions or slurries to white or grey cement concrete. Purer colours are obtained when coloured pigments are added to white cement concrete. Typical dosages are in the range 1–10% by mass of cement. It may be required to include a water-repellent admixture to assist colour maintenance [53]. Further advice is obtainable from Concrete Society Current Practice Sheet No. 99 [53].

It is essential for batching and mixing equipment and delivery vehicles to be cleared of material which might affect colour or shade adversely. This can create difficulties when concrete supplies are small or intermittent and when insufficient notice is given. Clients are advised to discuss their needs with ready-mixed concrete suppliers well in advance.

1.3.7 *Foaming agents*

Foamed concrete mixes have been developed for reinstatement of openings in highways and footways. The procedure involves mixing a foaming agent and a fine aggregate concrete, which can flow and self-compact in narrow pipe or cable trenches.

1.4 Other materials

1.4.1 *Pfa to BS 3892: Part 2, Grade A*

Pfa to BS 3892: Part 2, Grade A, is permitted by BS 5328 to be used in concrete, but may not be included as part of the minimum cement content for durability. It can be expected to reduce the water demand and increase later strength, but to a lesser extent than a Part 1 pfa.

It is particularly useful for overcoming deficiencies in fine aggregate grading, acting as a filler in lean mixes and as a pumping aid.

1.4.2 *Silica fume (micro-silica)*

Silica fume or micro-silica is a byproduct of the manufacture of silicon or various silicon alloys such as ferrosilicon. It is not yet covered by a British Standard and is not recognized by concrete standards or codes, but one source and type imported into the UK is covered by an Agrément Board Certificate Different sources and types can be expected to behave differently [54].

Silica fume is an ultra-fine material having pozzolanic properties, with a particle size of about 0.15 micron and a surface area of about $15\,000\,m^2/kg$. Its relative density is 2.20, similar to pfa. It is claimed to be extremely effective in increasing 28-day strength and in reducing risk of damage through ASR. For this latter benefit, it has been used effectively in Canada, Norway and Iceland [55].

The typical addition rate in the UK is 8% by mass of cement. Above this rate, significant increase in water demand may occur, requiring the use of a plasticizer or superplasticizer [56, 57]. The material available in the UK is normally provided as a 50:50 silica fume:water slurry for ease of handling and to avoid dust nuisance.

Up to 40% increase in strength has been observed when 8% silica fume has been incorporated in concrete, the benefit being significantly less at high cement contents [58]. At the time of writing, silica fume is not permitted by standards or QSRMC to be treated as part of any specified minimum cement content for durability. The fineness of silica fume can be of obvious benefit in reducing bleeding and segregation and in improving pumpability of concrete [55].

1.4.3 Fibres

Steel or polypropylene fibres have been incorporated into ready-mixed concrete with the purpose of increasing the tensile strength or impact strength of concrete. It is important to ensure that loose fibres are well spread through the concrete during charging of the mixer to avoid agglomeration [59]. Distributing fibres on belt-fed aggregates has been advocated. In some cases, it may be necessary for the fibres to be introduced during remixing of concrete at the site. Quoted concentrations are 0.5 to 1% by volume of steel fibres and 0.4% for polypropylene fibres [60].

Conventional workability tests may be used for concrete prior to addition of fibres, or appropriate limits chosen by trial for acceptance testing of fibre concretes.

1.5 Water for concrete

Typically, half the water requirement of concrete is present as free moisture in the aggregates. In the case of very low workability concrete suitable for rolling, it is important to ensure that the aggregates do not contain more water than that needed by the concrete; this may be only 5–7% by mass of the aggregates.

Water fit for drinking (potable water) is accepted as suitable for adding to concrete to supplement that in the aggregates, but other sources may be equally acceptable. BS 3148 provides a test method by which a proposed water may be assessed against a deionized water. The basis is comparison of concretes made with the two waters.

It is possible for normally suitable ground water or stream water to become contaminated, and periodic checks may be necessary. This may apply equally to the water used for washing the aggregates.

2 Properties of fresh concrete

Fresh concrete is concrete in a mouldable condition, able to be placed, compacted and finished by the chosen means. The properties of fresh concrete [61] which are of interest to the ready-mixed concrete producers and their clients are uniformity, stability, workability, pumpability, water demand and water/cement ratio, rate of change of workability, and finishability.

2.1 Uniformity and stability

Fresh concrete, properly designed, batched and mixed, needs to be sensibly uniform and to remain reasonably so through the remaining normal processes of transporting, placing, compacting, finishing and thereafter, until the stiffening process of hydration provides the necessary rigidity. Extreme construction conditions may require additional precautions. The range of relative densities of the normal materials used, 1 for water to over 3 for cement, means that there is always a tendency for segregation to take place, particularly at high workability. This tendency will generally be smaller for richer mixes, finer cementitious materials, finer sands, higher sand contents, lower water demand concretes, in warm weather, and when air-entraining and certain water-reducing admixtures are used. Agitation during transport or remixing before discharge of medium- and high-workability concrete may be necessary to maintain uniformity, whereas for low-workability mixes, agitation may be detrimental, causing aggregation or 'balling' of the fine components.

High sand contents can produce a problem during compaction and finishing, leading to the formation of a mortar layer at the surface if the concrete is over-vibrated, over-tamped or over-trowelled. Very high sand contents may also lead to a higher water content.

Some methods of handling during placing, such as dropping concrete through a cage of reinforcing bars, are likely to produce segregation of coarser from finer components. The user needs to appreciate that well-designed concrete will not overcome bad construction techniques.

Some bleeding of water is to be expected for most concretes, but excessive bleeding can result in significant settlement of the solid materials, and lead to cracking of the surface of concrete slabs over reinforcing bars and to 'sand-runs' on the vertical surfaces of walls or columns. Sometimes, the user of concrete takes into account the structural needs of the specification but ignores consideration of the fresh concrete. The cement content required for adequacy of finish can often be significantly higher than that needed for structural strength alone.

It is important for users to discuss with ready-mixed concrete producers any special construction situation so that any modifications to mixes can be discussed and agreed to combat adverse circumstances.

2.2 Workability

The term workability covers a wide range of properties [62]:

Mobility   (ability for concrete to move around reinforcement and into restricted spaces)
Compactibility   (by hand or vibration)
Finishability   (of free or moulded surfaces)
Pumpability   (for pumped concrete).

In any given construction situation, a given concrete may perform well, while in another situation it may perform less well. For example, a 40 mm maximum size aggregate concrete may be much easier to vibrate than a 10 mm aggregate concrete in a mass pour, but 10 mm concrete may be better in a narrow reinforced construction.

Greater importance may attach to different aspects of workability depending on the type of construction:

Mass pour mCf
Thin slab, lightly reinforced mcF
Intricately shaped structural section MCF

where m, c and f refer to the needs for mobility, compactability and finishability and the size of letter refers to its relative importance.

It follows that no one workability test method can ever adequately suit all construction situations. This is why it is important that the selection of workability and test method is left by the specifier to the decision of the contractor on the basis of his skill and experience of the type of construction and the equipment he elects to use to place, compact and finish the concrete.

However, it is vital that the contractor decides upon the workability needed prior to the stage of enquiry, because the quotation from the ready-mixed concrete supplier will be based upon the notified value. If at the order stage, for a designed mix, the contractor on site orders a higher workability, the ready-mixed concrete supplier will need to adjust the mix from that which formed the basis of quotation and adjust the price correspondingly.

2.2.1 Workability test methods

There are two reasons for the contractor specifying a workability method together with an appropriate target value, and invoking corresponding compliance limits. They are:

(i) To control workability within a selected range appropriate to the construction conditions and equipment to be used for transporting, placing and compacting concrete

(ii) To control indirectly the water content and thus water/cement ratio of the concrete.

The commonest method of test for both purposes is that for slump. Compacting factor is sometimes used for paving-quality concrete and the flow test for self-compacting or flowing concrete. The Vebe test and Tattersall 2-point test [63] are sometimes used in research laboratories, where they serve an important function in leading to a greater understanding of the mechanisms which affect workability, but they are only rarely used in the field.

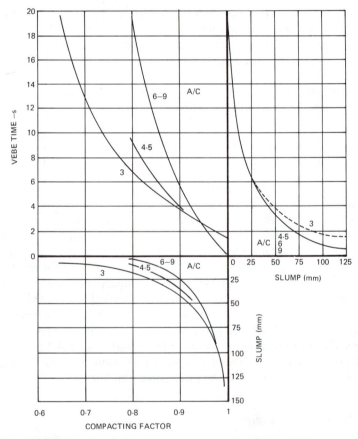

Figure 2.1 Summary of relations between slump, compacting factor and Vebe tests indicating the effect of aggregate/cement ratio. Redrawn and adapted from Dewar [64]

Table 2.1 Compliance requirements of BS 5328 for slump

Method of sampling	Specified slump	Tolerance
Representative sampling	All values	± 25 mm or $\frac{1}{3}$ of the specified value whichever is the greater
Spot sampling (see Fig. 7.1)	25 mm	$+ 35$ mm $- 25$ mm
	50 mm	± 35 mm
	75 mm and over	$\pm$ ($\frac{1}{3}$ of specified slump plus 10 mm)

The relationships between the different methods are dependent upon mix characteristics. Figure 2.1 shows how relationships are affected by aggregate/cement ratio for slump, compacting factor and Vebe time [64].

The compliance requirements in BS 5328 for the common workability tests are shown in Tables 2.1 and 2.2.

BS 5328 recognizes the practical difficulty that, in order to check concrete for compliance for slump, it is necessary to obtain a representative sample which would entail discharge of the complete delivery. To overcome this problem, spot sampling (see Fig. 7.1) from an early part of the discharge, after $0.3 \, \text{m}^3$ has been discharged, is permitted for the slump test only, but with increased tolerance. For compacting factor, for cube-making and for other tests, samples are required always to be fully representative, entailing sampling through the complete discharge (see Fig. 7.1).

The commonest specified slumps are 50 mm and 75 mm, both with tolerances of ± 25 mm. The acceptable range of 25–75 mm for a specified slump of 50 mm is deceptively large. The slump test is so sensitive to variation in water content that for concrete to be within the allowed slump tolerance it will need to be controlled to within -5% and $+3\%$ of the target water content or water/cement ratio, which is very tight control indeed. At 75 mm slump the control is even tighter, -3 to $+2\%$. Thus, the slump test and its corresponding limits are excellent and longstanding ways of controlling workability and water content of concrete, although, regrettably, they are not always recognized as such.

Tables 2.3 and 2.4 provide guides for the contractor in the selection of slump values for different purposes. Guidance in Table 2.3 is similar to that given in BS 5328.

Table 2.2 Compliance requirements of BS 5328 for compacting factor

Method of sampling	Specified compacting factor	Tolerance
Representative sampling only	0.90 or greater	± 0.03
	0.81 to 0.89	± 0.04
	0.80 or less	± 0.05

Table 2.3 Selection of slump for different construction situations [65]

Degree of workability	Nominal specified slump (mm)	Tolerances BS 1881 sample (mm)	Tolerances BS 5328 alternative method (mm)	Form of compaction	Use of concrete
Very low	10	0–25	0–35	Heavy vibration	Roads laid by power operated machines.
				Light tamping	Kerb bedding and backing
Low	50	25–75	15–85	Poker or beam vibration	Power floated floors. Vacuum processed floors and roads.
Medium	75	50–100	40–110	Poker or beam vibration and/or light tamping	Strip footings. Mass concrete foundations. Blinding. Normal reinforced concrete in slab beams, walls and columns. Sliding formwork construction. Pumped concrete. Vacuum processed concrete. Domestic general purpose concrete.
High	125	80–165	70–175	Self-compaction Poker vibration	Trench fill. In-situ piling*. Concrete section containing congested reinforcement.
Very high	150	100–200	90–210	Self-levelling	Diaphragm walling*. Self-levelling superplasticized concrete.

* See also Table 2.4.

2.3 Pumpability

The ability of concrete to be pumped [67–69] relies on a combination of:

(i) Properties of the concrete
(ii) Pipeline diameter and capability of the pump

Table 2.4 Selection of slump of flow values for cast-in-place piling.

Typical conditions	Target slump	Target flow
Poured into water-free unlined bore. Widely spaced reinforcement.	100, 125	—
Where reinforcement is closely spaced. Where casting level is within the casing. Pile dia less than 600 mm.	125, 150	400, 450
Where concrete is to be placed by tremie under water or drilling mud	175	500
Pumped concrete for CFA and small diameter piles.	200	550

Note: Slump and flow values do not necessarily correspond to each other. The suitability of these values needs discussion between contractor and concrete supplier to take account of site conditions and piling techniques.

(iii) Construction situation – height of outlet; distance to be pumped; number and angle of bends; continuity and rate of pumping.

As far as the concrete is concerned the prime properties are:

(i) Ratio of max aggregate size to pipeline diameter
(ii) Workability
(iii) Stability.

Generally, concrete will normally be pumpable if the following apply:

(i) Cement content is over $300 \, kg/m^3$ or the weight of cement plus fine aggregate below 300 micron is over $1.75 \times$ free water content.
(ii) Slump is over 50 mm, preferably target 75 mm
(iii) Nominal maximum aggregate size not exceeding one-fifth of the pipeline diameter, i.e. 20 mm for a 100 mm pipeline
(iv) Fine aggregate content is up to 5% higher than for a normal well-designed mix.

Concretes which bleed badly may be unpumpable. This may be overcome by adding more cement, pfa, finer sand or some water-reducing agents. Excessively cohesive concretes, particularly high cement content mixes, may be unpumpable due to excessive friction. Increasing slump may help in this case. It may be necessary to reduce 10 mm content in a 20 mm aggregate concrete, especially if the 10 mm material is flaky.

2.4 Water demand and water/cement ratio

The initial water demand of concrete is influenced by:

(i) Cementitious materials characteristics and content

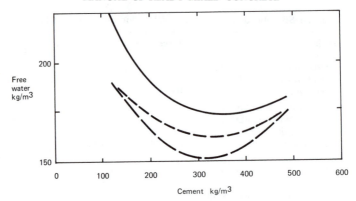

Figure 2.2 Examples of relations between water content and cement content for three different materials combinations in concrete of 50 mm slump. From Dewar [12]

(ii) Aggregate characteristics
(iii) Aggregate moisture condition and absorbency
(iv) Required workability – specified workability plus allowance for loss of workability in transit
(v) Ambient conditions.

Typical free water demands lie in the range 150–220 litres/m^3 for medium workability concrete.

The water demand varies with cement content as shown in Fig. 2.2, the minimum value occurring at some intermediate cement content dependent upon cement and aggregate characteristics.

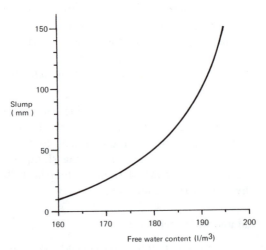

Figure 2.3 Example of relationship between slump and water content. Based on data from Dewar [61]

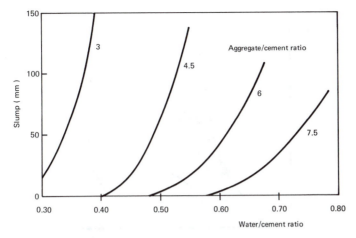

Figure 2.4 Example of effect of aggregate/cement ratio on relationship between slump and water/cement ratio. After Newman [62]

No particular technical significance is attached to the minimum value and its corresponding cement content, because cement content will in all probability be determined by considerations of strength and durability. However, it is possibly significant that the normal working range for cement content does lie around that producing the lowest water demand.

Water demand is only slightly affected by the ratio of fine to coarse aggregate, but can be significantly increased when the ratio is high compared with the optimum [12].

Higher temperatures of the environment and the materials increase the water demand of the concrete, but the net effect on cement content for a given strength and slump is unlikely to be very great in the UK under normal conditions. A 17 °C temperature rise required 2–6% increase in water, but less than 5 kg/m³ increase in cement content in recent US tests [70].

A relation between workability and water content is summarized in Fig. 2.3, and is sensibly independent of cement content [61]. However, the relationship between water/cement ratio and workability is influenced by cement content [62] as is shown in Fig. 2.4.

The influence of time and transporting are considered in detail in the following sections.

2.5 Rate of change of workability

All concretes can be expected to lose workability slowly with time due to effects of hydration, associated temperature rise, evaporation, and, in the case of dry aggregates, absorption of water [61, 71, 72]. These effects will be much magnified in the laboratory, so that totally misleading information on loss of workability with time can be obtained.

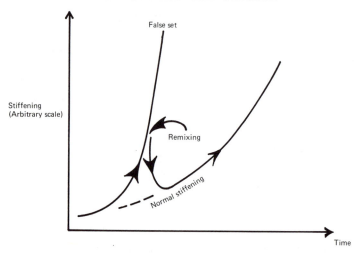

Figure 2.5   Illustration of effect of remixing on workability of concrete which has suffered 'false setting'

There are certain situations in real practice which can result in fast or flash stiffening of concrete:

(i) Extreme hot dry weather
(ii) Use of very hot water for mixing in winter
(iii) Excessive contents of accelerator, particularly calcium chloride (now banned for use in structural concrete)
(iv) Use of dry aggregates (particularly lightweight absorptive aggregate)
(v) Cements with more rapid setting times (see also section 1.2)
(vi) Flash-setting cements (see also section 1.2)
(vii) False-setting cements (see also section 1.2).

In the case of the false-setting cement, there is a fast loss of workability which can be regained if the concrete is remixed or vibrated [61] as shown diagrammatically in Fig. 2.5.

Provided the concrete is in the process of being mixed or can be remixed, which is normally not a problem with agitated ready-mixed concrete, the workability is quickly regained. When false-set occurs with lean or low workability concrete transported in a truckmixer without agitation or by a tipper, then it can create a problem which can only be overcome by having sufficient drive power to commence agitation or by intense vibration during placing and compaction [61].

2.6   Influence of transporting on the workability of ready-mixed concrete

Ready-mixed concrete will be affected not only by time but by any factors associated with transporting. The effects of transporting may depend upon

Table 2.5 Mixing and transporting of ready-mixed concrete [61].

Method	Mixing at the plant	Transporting container	Mixing at site
1	Mixing completely in a plant or truckmixer	Truckmixer drum revolving at agitating speed	Remixed for short time only
2	Mixed partially in a plant or truckmixer using proportion of water only	Truckmixer drum but not revolving	Remainder of water added and concrete remixed for several minutes
3(a)	Mixed completely in a plant or truckmixer	Truckmixer drum but not revolving	Remixed for short time only
(b)	Mixed completely in a plant mixer	Tipper	None

whether the concrete is agitated in transit and whether it is remixed before discharge (see Table 2.5) [61].

Method 1 (Table 2.5) requires the plant supervisor to anticipate the normally small effects in transit by adjustment of water content, and allows the workability and water content of all deliveries to be controlled by one experienced operative. Method 2 is preferred by some ready-mixed concrete companies, and requires water content to be adjusted and workability to be judged by each truck driver at delivery. Method 3 is used for dry mixes, particularly dry-lean concrete, 3(b) being used commonly by customers employing their own transport to collect low-workability concrete.

During transport, non-agitated concrete will tend to settle, compact and to segregate, requiring expenditure of energy in remixing to return it to a loose workable state and to regain uniformity. An exception would be low-workability lean concrete which would be relatively unaffected (but see false setting, section 1.2.4).

Agitation of concrete has the important function of counteracting segregation and compaction during transport, and thus maintains uniformity and the bulked, workable state to the point of delivery. Airflow during agitation might increase evaporation, leading to a reduction in workability. The friction caused by agitation could result in grinding, which in turn might lead to an increased water demand and thus reduced workability. It is assumed that heat developed during grinding is compensated for by the cooling effect of evaporation.

The action and interaction of all these factors, together with those described in the previous section dealing with the influence of time, are summarized in Fig. 2.6.

The four major influences affecting workability are evaporation, hydration, absorption and grinding. Of these, probably both evaporation and hydration accelerate with time, depending upon ambient conditions. Absorption or grinding are likely to produce significant effects only when exceptionally

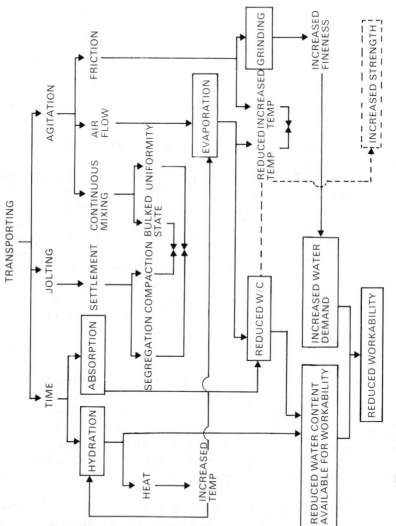

Figure 2.6 Interaction between the various factors involved in transporting ready-mixed concrete and their effects on workability and strength. After Dewar [61]

Table 2.6 Internal influences on the effects of transporting concrete [61].

Materials	Temperature
Cement type	Early hydration characteristics
	Fineness
	Heat development characteristics
Cement content	Heat development
	Water utilization in hydration
Aggregate	Friability and abrasive resistance
	Moisture condition
	Porosity
Water content	?
Initial workability	Influence on grinding
	Influence on hydration
Admixtures	Influence on hydration
	Influence on heat development
	Influence on air content

absorptive and dry aggregates or abradable aggregates are used.

Another possible factor is grinding of cement particles [73]. It is not known to what extent this occurs and whether the effects are significant, but grinding of cement particles would tend to increase the water demand of the fresh concretes of medium- and high cement contents and thus reduce their workability. It may even be possible in a lean mix, due to shortage of fine particles, for grinding of cement to have the reverse effect by reducing water demand, and for an increased workability to be observed.

The effects of the main factors in reducing workability can be considered in three separate ways associated with the water content of the concrete:

(i) Increasing the water demand, e.g. by increased fineness
(ii) Extracting water from the paste by evaporation or absorption, resulting in a reduced effective water/cement ratio
(iii) Decreasing availability of water in the paste for workability, e.g. by hydration, without changing the water/cement ratio.

Due to the interaction of the various influences, the rate of reduction in

Table 2.7 External influences on the effects of transporting concrete [61].

Ambient conditions	Temperature, humidity and wind speed
Transporting method	See Table 2.5
Transporting container	Volume
	Air access
	Material and colour (heat radiation)
Bulk volume of concrete	
Journey time	Road distance, road and traffic conditions, gradients
	Site access
Discharge	Waiting time
	Rate of pour

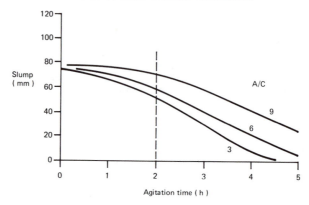

Figure 2.7   Effect of aggregate/cement ratio on rate of loss of workability with time. After Dewar [61]

workability increases with time, generally, but may not produce losses of practical significance until several hours after the initial mixing.

This rate of reduction in workability is influenced by internal and external conditions, as indicated in Tables 2.6 and 2.7. Some of these influences are discussed in more detail in the remainder of this section.

2.6.1 Cement content

Mixes of lower cement content lose workability at a lower rate because the rate of temperature rise is less and a smaller proportion of water is utilized in the hydration process during a given time. This is illustrated in Fig. 2.7.

2.6.2 Aggregate grinding

Some aggregates, particularly limestone fine aggregate or weakly cemented sandstone, are likely to be abraded slightly during agitation, resulting in increased fineness and higher water demand. Where necessary, this could be offset by reducing slightly the initial fine/total aggregate ratio of the mix.

2.6.3 Water content and initial workability

Mixes of higher water content or initial workability lose workability at a lower rate because the effect of a given water loss, due to evaporation or hydration, will be diluted. This is illustrated in Figs. 2.8 and 2.9.

2.6.4 Admixtures

Admixtures may influence the heat development, hydration or air content and may thus influence the rate of loss of workability. In particular, calcium

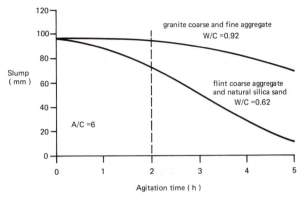

Figure 2.8 Effect of water/cement ratio on rate of loss of workability with time. After Dewar [61]

chloride (rarely permitted) may increase appreciably the rate of hydration and temperature rise, leading to rapid stiffening, particularly when rich dry mixes are transported in hot weather. Because the early rate of loss of workability is usually due primarily to evaporation, retarders may not have a significant effect until a later stage when hydration becomes dominant. Thus retarders will tend to influence the later rate but not the early rate of loss of workability. The air content of air-entrained concrete may also change with time, resulting in increased or decreased workability.

2.6.5 Ambient conditions

Low air or materials temperatures decrease the rates of hydration and evaporation, and thus decrease the rate of loss of workability. The normal range of temperature of cement is unlikely to cause significant effects, but very

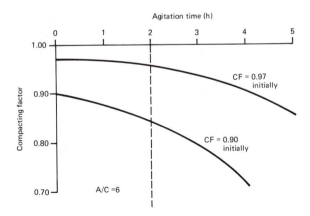

Figure 2.9 Effect of initial value on rate of loss of workability [61]

high values may contribute to problems with rich mixes in hot weather. Obviously, high humidity of the air decreases the rate of evaporation and consequent loss of workability.

Wind speed possibly has only a small influence on protected concrete in a drum, but becomes more important during discharge and placing, when the concrete is unprotected.

2.6.6 *Bulk volume of concrete*

Larger volumes of concrete have a lower surface area/volume ratio and are less susceptible to large losses in water due to evaporation or to loss of a significant proportion of mortar through adherence to the drum or its blades. On the other hand, large volumes retain heat more easily and may produce a higher rate of hydration. It is not known how all these aspects interact in detail with each other to affect loss in workability, but practical experience suggests that large volumes lose workability at a significantly lower rate than small volumes.

2.6.7 *Transporting method*

It is probable that different transporting methods will produce different effects but it seems unlikely that Methods 1 and 2 (see Table 2.5) will produce effects which are seriously different, because a long period of low-speed agitation may be considered to be equivalent in its effects to a short period of high-speed mixing.

Obviously, uncovered concrete transported by Method 3(b) may suffer greater effects at the surface due to evaporation, but is protected internally.

2.7 Effects of transporting of concrete on compressive strength and workability

To assist with discussions on limitations of delivery time and retempering, it is essential to consider the interrelated effects of transporting on strength and workability. Strength has been found to increase generally with time at about 5% per hour of agitation, if the concrete can be compacted fully at the decreased workability [71].

Figure 2.6 indicates that, when damp aggregates are used, concrete strength increases only through evaporation, but that workability decreases because of other effects as well. Thus, the relation between strength and workability for a given mix, as initial water content is altered, can be expected to change with time of agitation, as confirmed by actual test results summarized in Fig. 2.10.

This effect is generally quite small for typical transporting times and conditions in the UK but becomes more important for very long periods of agitation, rich mixes, low initial workabilities and in hot dry weather.

In practice, the effects even though small are taken into account in the design

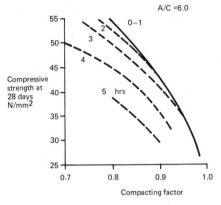

Figure 2.10 Effect of agitation time on the relation between strength and workability. After Dewar [61,71]

of concrete because design data are usually based on strength and workability tests made at delivery and not at the plant.

2.8 Limitations on delivery time

Misconceptions concerning the mechanisms of setting or stiffening have caused fears that there is a particular time after which it is detrimental to continue working fresh concrete. Technical education, particularly by the British Ready Mixed Concrete Association and Cement and Concrete Association, has now largely overcome this problem.

It is true that for each mix and particular means of compaction there will be a time after which concrete will be losing workability at too great a rate to allow full compaction to be obtained. This limiting time will depend on the mix and on the external conditions, and will be longer for lean, high-workability mixes, in cold weather and when vibration is employed.

There will also be a time, depending again on the mix and conditions, when

Table 2.8 Approximate times of agitation after which the rate of loss in compacting factor exceeds 0.05 per hour, under average conditions.

Aggregate/cement ratio (by weight)	Agitation time (h)	
	Initial slump (mm)	
	25	125
3	1	2
4.5	1.5	3
6	2.5	4
9	3.5	5

Table 2.9 Approximate times of agitation at which a strength reduction of $2\,N/mm^2$ occurs below that for concrete of the same slump but compacted shortly after mixing.

Aggregate/cement ratio (by weight)	Agitation time (h)	
	Initial slump (mm)	
	25	125
3	1	2
4.5	2	3
6	3	4
9	4	over 5

the strength obtained at the specified workability is significantly lower than that obtained for concrete placed immediately after mixing. However, it is usually impractical to specify different time limits for different mixes and conditions.

Some specifications provide a single time limit, covering all mixes and conditions, which is easy to comprehend and enforce and is intended to allow the contractor sufficient time subsequently to place and compact the delivered concrete without appreciable change in workability occurring. If the time limit is not unduly restrictive, problems will not usually occur for the producer. BS 5328, for example, requires ready-mixed concrete to be delivered within two hours of mixing when agitated, or one hour if not agitated.

The reasonableness of this simple approach can be assessed for average ambient conditions from Tables 2.8 and 2.9. For example, if a time limit of two hours is specified then, from Table 2.8, most concretes delivered at the specified workability within this time would not be losing workability at an excessive rate and, from Table 2.9, no excessive strength loss would be expected compared with concrete of the specified slump placed soon after mixing.

Again, it is worth stressing that when concrete quality control is based on samples taken from delivery, any effects of time on workability or strength are accounted for automatically. In any event, in the UK it is very rare for delivery times to approach the maximum allowed time of two hours.

2.9 Retempering of concrete

Water properly mixed with fresh concrete at some time after the initial mixing behaves as if it had been added initially [71]. Thus, retempering could be used to restore workability to the specified value, provided the water can be mixed intimately with the mass of concrete. This will normally be possible only when a truckmixer is used to transport the concrete.

When concrete is observed at delivery to have a lower workability than

Table 2.10 Approximate times of agitation after which the relations between strength and workability commence to deviate significantly from those for concrete placed shortly after mixing

Aggregate/cement ratio (by weight)	Agitation time (h)	
	Initial slump (mm)	
	25	125
3	0.5	1
4.5	1	2
6	2	3
9	3	4

specified, there are three possible reasons:

(i) Insufficient water batched initially
(ii) Higher rate of evaporation (or absorption) than anticipated
(iii) Higher rate of hydration (or grinding) than expected.

Retempering water added to offset (i) or (ii) will not result in a lower strength than expected, whereas extra water to combat (iii) will result in a lower strength. However, in examining concrete at delivery it will not be possible to determine which reason applies.

Under average conditions, hydration should only start to produce significant effects after the times shown in Tables 2.10, so that retempering for any of the three reasons could be carried out with reasonable confidence within these times. If a slight loss in strength is acceptable due to (iii), the times could be relaxed to those given in Table 2.9, so that for nearly all mixes retempering within the normally specified maximum delivery time of two hours would be acceptable.

As an extra safeguard, it might be advisable for the producer to limit the quantity of retempering water to some maximum value, say, 6 litres/m^3, which is approximately 3% of the total mixing water requirement. This would then limit any strength reduction to less than 2 N/mm^2. From Fig. 2.3, retempering of 25 mm slump concrete with 6 litres of water per m^3 would allow the slump to increase to 35–40 mm, well inside the tolerance of 25–75 mm for 50 mm specified slump.

Some confirmation of the reasonableness of this approach is provided by Fig. 2.11, where ready-mixed concrete has been retempered with extra water for two hours to maintain slump, without strength dropping below the initial level [131].

Policy within the ready-mixed concrete industry varies regarding retempering. Some companies, particularly those requiring all water to be added at the plant, do not allow concrete to be retempered unless the extra water is signed for by the customer's representative. This has the advantage of protecting the

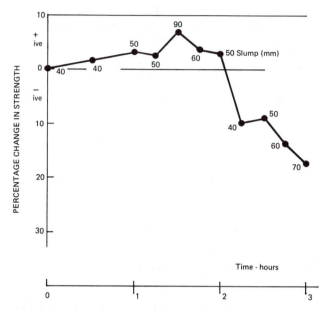

Figure 2.11 Effect on strength of retempering concrete with additional water to maintain approximately 50 mm slump for up to 3 h. Redrawn and adapted from Beaufait [131]

customer by restricting the unauthorized addition of extra water to produce a higher workability than specified. For other companies, when water is always added at site it is not possible to distinguish between mixing water and retempering water since they are added at the same time. It is, however, possible to restrict the total amount of water added to some fixed amount, and this is usually done; any extra will again require a signature from the customer. BS 5328 requires that all additional water is recorded on the delivery ticket.

2.10 Laboratory simulation of ready-mixed concrete

Difficulties occur in attempting to simulate ready-mixed concrete through laboratory tests. Small-scale experiments may reproduce faithfully the qualitative effects, but may overestimate grossly the quantitative effects of mixing, time or agitation [61].

Regrettably, much research intended to simulate ready-mixed concrete has been wasted by not taking this point into account. During the first few hours, when using dry aggregates in laboratory mixes, most loss of workability is due to absorption and evaporation, coupled with loss of moisture and fine materials on mixing and testing equipment.

The following factors may need to be taken into account in the planning of small-scale experiments and in the interpretation of data and conclusions from past research in the laboratory relating to ready-mixed concrete:

(i) Volume of concrete
(ii) Exposed surface area/volume ratio
(iii) Mixer/agitator power, rpm and mixing action
(iv) Water absorption by dry aggregates
(v) Water absorption, evaporation and loss of mortar in coating the drum and blades and the sampling and testing equipment
(vi) Abrasion or crushing of aggregates dependent on mixer power and blade configuration.

To ensure validity of conclusions in respect of mixing, agitation or time, it is essential that tests are made on full-scale mixes using normal production equipment and conditions.

3 Properties of hardened concrete

3.1 Surface quality

The surface quality achieved is a perfect mirror of the combined effects of concrete specification, concrete quality provided, the quality of the formwork (in the case of vertical faces and soffits) and the qualities of supervision and workmanship in placing, compacting, curing and finishing of the concrete.

Commonly specifications for concrete concentrate on strength and durability, ignoring surface finish. C & CA advice for high standard finish of concrete includes a recommendation for a minimum cement content of $320 \, kg/m^3$ [74]. An earlier more detailed C & CA publication advised a cement content between $355 \, kg/m^3$ and $475 \, kg/m^3$ for $20 \, mm$ aggregate concrete [75].

So-called sand-runs (or scouring) on vertical surfaces are to be expected when lean mixes, low grades and high workabilities are specified. The problem will be increased in cold weather and with deep lifts. The problem is caused by bleeding of water up the face of the formwork. It may be difficult to minimize the occurrence of surface blowholes when formwork is impermeable, when concrete is air-entrained or when rich mixes are employed. Comparing the different advice given in relation to blowholes and sand-runs leads to the conclusion that minimizing the risk of one fault will increase the risk of the other. This needs to be taken into account in selecting a solution. When colour variation must be minimized, ensuring that materials are always from a single source will usually be necessary, to complement the many aspects of site construction which must be under strict control.

Detailed advice on diagnosis and control of blemishes is provided in C & CA *Appearance Matters* No. 3 [76].

3.2 Cracking of concrete

All concretes crack, indeed if they did not, there would be something wrong, because the mechanisms of cracking and creep can be thought of as important means by which stresses in a heterogeneous material readjust themselves. Fortunately, most of the cracking is on a microscopic scale and does not constitute a normally visible fault.

Concrete, like the human body, can tolerate some degree of damage and is also self-healing. The mechanism, known also as autogeneous healing, permits hydration products to form in water-filled cracks, so sealing them. As a result,

Table 3.1 Classification of non-structural cracks. Modified from [77]

Type of cracking	Time of appearance	Letter (see Fig. 3.1)	Subdivision	Most common location	Primary cause (excluding restraint)	Secondary causes/ factors	Preventive measures
Plastic settlement	Ten minutes to three hours	A	Over Reinforcement	Deep sections	Excess bleeding	Rapid early drying conditions	Reduce bleeding e.g. Air entrainment. Revibration.
		B	Arching	Top of columns			
		C	Change of depth	Trough and waffle slabs			
Plastic shrinkage	Thirty minutes to six hours	D	Diagonal	Roads and slabs	Rapid early drying	Low rate of bleeding	Improve early curing.
		E	Random	Reinforced concrete slabs			
		F	Over reinforcement	Reinforced concrete slabs	Ditto plus steel near surface		
Early thermal contraction	One day to two or three weeks	G	External restraint	Thick walls	Excess heat generation	Rapid cooling. Inadequate distribution steel.	Reduce heat and/or insulate. Increase distribution steel
		H	Internal restraint	Thick slabs	Excess temperature gradients		
Long term drying shrinkage	Several weeks or months	I		Thin slabs (and walls)	Inefficient joints	Excess shrinkage. Inefficient curing. Inadequate distribution steel.	Reduce water content. Improve curing. Increase distribution steel.
Crazing	One to seven days sometimes much later	J	Against formwork	'Fair faced' concrete	Impermeable formwork	Rich mixes Poor curing.	Improve curing and finishing.
		K	Floated concrete	Slabs	Over-trowelling		
Corrosion reinforcement	More than two years	L	Natural	Columns and beams	Lack of cover	Low specified concrete grade. Poor curing. Poor compaction.	Increase cover. Higher grade. Good compaction. Good curing. Use nonchloride accelerator.
		M	Calcium chloride	Precast concrete	Excess calcium chloride		
Alkali–silica reaction	More than five years	N		(Damp locations)	High alkali cement and reactive aggregate		See advice on ASR

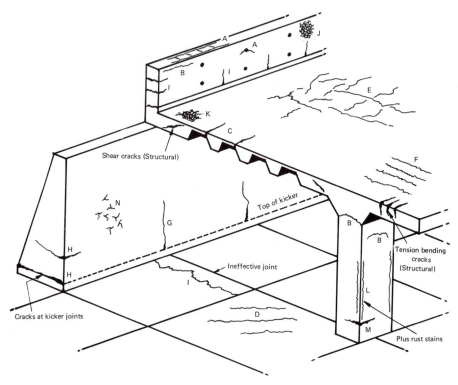

Figure 3.1 Examples of cracks in a hypothetical concrete structure. Redrawn and adapted from Concrete Society [77]

cracks in foundations and even in water-retaining structures can often be tolerated.

Some cracks, however, are not only visible but may be unacceptable either structurally, aesthetically or because of influences on durability or serviceability, and may require expensive remedial treatment.

Cracks may be initiated in plastic, stiffening or hardened concrete. Their causes could be many and various and they may be difficult to diagnose. Fig. 3.1 provides graphic guidance on the more common types of non-structural and structural cracks, and Table 3.1 provides a summary of the potential causes of non-structural cracks, together with preventive measures. For more detailed information on diagnosis and remedial measures, readers are recommended to consult the Concrete Society Report [77].

To avoid cracking of the surface of sloping slabs, it is important for compaction and finishing to be directed up the slope, which is against gravity and the natural inclination of the operative.

3.3 Strength

The strength of concrete is both a key structural attribute and also a measure of the development of hydration by which many other properties can also be expected to improve with time. Thus strength has become the most commonly specified and tested property of hardened concrete.

Of the various strengths which might be measured, the compressive strength of cubes made, cured and tested at 28 days in a standard way has become the accepted means by which quality is assessed and other properties estimated. Only very occasionally are other strengths actually measured, reliance being placed on empirical relationships, e.g. between compressive strength and tensile [78], flexural, shear or bond strengths. These strengths are typically an order of magnitude lower than compressive strength.

Periodically, attempts are made to promote tensile strength testing, but precision so far has been too poor for use for compliance purposes, as may be seen from Fig. 3.2.

Typical values for cube strength range up to $70 \, \text{N/mm}^2$; higher values are achieved only with selected cements and aggregates, by the use of water-reducing admixtures, by accepting lower workability or waiting beyond 28 days [79].

The range of relationships of strength with cement content or with water/cement ratio in the UK is quite wide for dense aggregate concrete, because of the range of cement and aggregate qualities which are in use, as may be seen from Fig. 3.3. The curves are based on survey data for 75 mm slump concrete obtained by BRMCA in 1980 [80, 81]. The flattening of the curves at high strengths is a function of the aggregate strength, cement capability and bonding of cement to aggregate, coupled with influences of both aggregate and cement on water demand for a stated workability in the

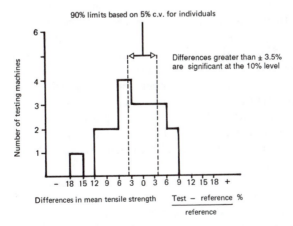

Figure 3.2 Distribution of differences between mean tensile strengths obtained on a range of commercial testing machines and results obtained on a reference machine. After Ryle [132]

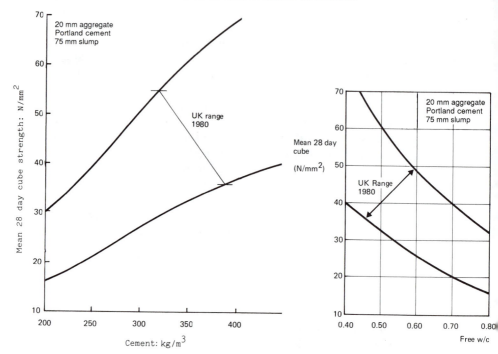

Figure 3.3 Range of relationships between (left) cement content and strength and (right) strength and water/cement ratio. After Dewar [80, 81]

case of the cement-content curves. As the particular 'ceiling' strength is approached, the benefit of increasing cement content is progressively reduced so that, eventually, further increase produces negligible benefit. In the middle of the range, $1 \, N/mm^2$ gain in strength may be obtained by increasing cement content by $4\text{--}8 \, kg/m^3$, depending on the particular materials combination.

Relationships between strength and cement content are very complex, and involve cement hydration and water demand chracteristics, proportions of the various components, bond between the cement hydration products and the aggregate particles, strength of the aggregate particles, strength of the cement hydration products, influence of admixtures, air content, compaction, curing and age. The relative importance of each factor varies with the strength level being considered.

There is a tendency for the rounder smooth aggregates to produce high strengths at low cement contents, because of the benefit to water demand, whereas the stronger crushed rocks may tend to produce the highest strengths at high cement contents. Thus it is possible for relationships for two aggregates with the same cement to intersect at some intermediate cement content.

3.3.1 *Strength development with age*

Under standard curing, concrete cubes develop strength very quickly, the greatest gain being observed in the first few days, with progressively less gain on each subsequent day, although gain may still be observed even after 10 years and more.

While this continued gain is reassuring to all concrete users, the length of time to achieve most of the ultimate strength would be rather long in contractual terms, so that an age of 28 days has emerged as a workable compromise for the basis of specification requirements. However, this is still too long for contractors and specifiers to wait before making decisions to remove forms, to stress concrete or to continue with further construction. Equally for the ready-mixed concrete producer, 28 days is too long to wait before taking action to remedy a production problem.

To overcome these problems, use is often made of relationships between early and 28-day strengths to enable early action to be taken. For example, contractors will often use 7-day strengths, and the ready-mixed concrete producer will use 7-day or even 1-day accelerated strengths [82–84], equivalent to 7 days' normal curing, in quality control systems.

It is important to know the relationships between early and 28-day strengths for the particular cement and aggregate combination in use, and to monitor the relationship. It is not sufficient merely to assume a value, to assume that the value determined at a medium level of strength applies at low or high levels, or that the value obtained from cement test data will necessarily be the same for the aggregates used in practice.

3.4 Durability

Concrete is a heterogeneous mixture of differing materials, some inert and some reacting, continually changing with age. This heterogeneity and mobility is at the same time both advantageous and a pitfall for the unwary. Concrete is amazingly versatile and capable of overcoming many of the misuses and mishandling to which it is subjected. Yet at the same time it is often misunderstood, abused and subjected to a lack of care which may fail to allow it to achieve anywhere near its full potential [85].

Concrete is more than a material, it is a complete process with changing control at different stages in the process. Specification to meet a need is the first stage and raw materials production is the second. The mix formulation, mixing of the fresh concrete and transporting to the construction form the third stage, and the placing, compaction around reinforcement and early curing make up the fourth. The fifth stage is maturing in the construction environment, the sixth is functioning under action of the working load and environment.

Each of the stages may involve different organizations and individuals who will influence subsequent performance in the final stage by their skills and care.

A lack of care in one stage can nullify all the care in each of the others, but when the media identify faults in concrete, it is concrete which is blamed, not that part of the concreting process which is at fault.

Where durability is concerned, the evidence of distress is that seen easily at the surface. This is usually because of two reasons, both self-evident. Firstly, only surface distress is easily observable, and secondly, most, although not all, durability problems start and end at or near the surface rather than at the heart. If we consider why this second observation is so common, again it should be self-evident that external aggressive agencies can more easily attack the surface than elsewhere. What may be slightly less evident is that the surface is where the lack of care in the fourth and fifth stages of concrete construction can wreak havoc. The surface is the Achilles heel of concrete [85].

3.4.1 Concrete mobility

Concrete is mobile with time, both chemically and physically, and in the fresh and hardened state. This mobility can be both helpful and harmful.

During and after compaction, air and water move upwards in the fresh and stiffening concrete. The benefit is a reduction in the porosity of the concrete. On the debit side, a water/cement ratio gradient develops. The permeability in the vertical direction can be increased by capillary channels which start as a network of tiny streams of water, developing into rivers in the heart and on vertical faces, into estuaries and even into an ocean at the top surface. This ocean, when it occurs, has a side benefit in providing some necessary water for curing but it is offset by a weakening of the surface. Of course, good concreting practice can minimize extreme effects but some effect will almost always occur.

Through sedimentation, which is the process of settlement of the fine particles of cement and sand, tiny flaws may develop under coarse aggregate particles and reinforcing bars.

In the stiffening and hardening concrete, continuous or spasmodic hydration helps to reduce permeability and seals cracks by autogenous healing. Hydration increases temperature. Thermal and moisture movements cause expansion, shrinkage, tensile, compressive and differential stresses, creep and cracking.

Despite all these events, concrete is an accommodating material, adapting generally very well, almost in spite of whether the mechanisms and the need for care are understood. Most of the flaws will be at the microscopic level and very few create obvious visible defects, but it is important not to be lulled by this into a false sense of security. Aggressive agencies can identify these weaknesses and take advantage of them.

In the hardened state, movement continues. Air, water vapour and fluids may enter and move around in concrete, evaporating water, depositing or dissolving chemicals or diluting solutions and relocating chemicals. In reinforced concrete, concrete variations, moisture variations, chemical

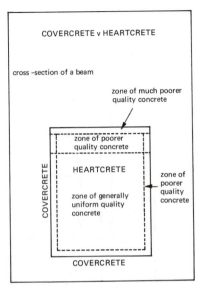

Figure 3.4 Effects of water gain and lack of curing leading to different qualities of concrete in an rc beam

migration and carbonation will change the electropotential distribution, which in turn may encourage electrochemical reactivity, ideal for corrosion. Loading, wind speed, daily and annual temperature changes, humidity changes, ground movements and water levels will all change the stress and strain concentrations.

Concrete can have a different permeability horizontally compared with vertically because of the effects of sedimentation and microscopic cracking. Compaction may be variable, due to the difficulty of compacting concrete well at edges and corners and near boxed-out areas, and between reinforcement and the surface. Concrete can have cracks in the heart which are invisible at the surface. Cracks at the surface may not progress far into the concrete, or they may go right through.

3.4.2 *Covercrete and heartcrete*

As a result of the mobile nature of concrete, variations in the concrete itself, and variations in compaction and curing, concrete construction will display a variation in properties over its surface and its cross-section, as shown, for example, in Fig. 3.4.

In order of quality, 'heartcrete' will usually be the strongest, and the surfaces, more particularly the top surface, will usually be the weakest. The differences can be very large. For durability it is now thought best to concentrate on the 'cover'. Durability is only skin deep, because of the location of the reinforcement near this cover layer (Fig. 3.5).

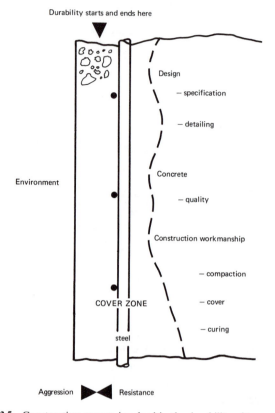

Figure 3.5   Construction aspects involved in the durability of 'covercrete'

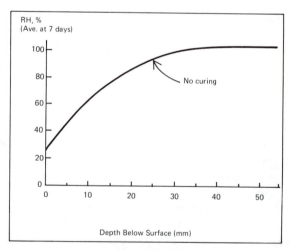

Figure 3.6   Example of the effect of poor curing on the humidity gradient within concrete at an age of 7 days. Redrawn and adapted from Spears [86]

The outer 25 mm or so may be an order of magnitude more permeable than the heart, due to poor protection and lack of curing. The upper 20% of the height may have a substantially greater permeability due to water gain (see section 3.4.1).

The influence of poor curing on the humidity of covercrete can be judged from Fig. 3.6, from an American report [86], which indicates that, even as early as 7 days, as much as 15 mm of the cover depth may be below 80% relative humidity, the minimum level accepted as necessary for continued hydration.

3.4.3 *A background to specifying durability*

It is rarely possible for specifiers to draft or for producers to accept specifications which refer directly to durability on a performance basis. The reasons are, firstly, that there are few recognized tests that can be used as a basis, and, secondly, that the concrete producer cannot easily take account of the unknown factors associated with workmanship and conditions of service under which the concrete will be expected to function and over which the concrete producer has no influence.

The commonest ways used by specifiers to overcome this problem are to specify designed mixes with one or more of the following requirements:

   (i) Strength grades selected for durability
  (ii) Minimum target cement content
 (iii) Maximum target water/cement ratio
 (iv) Maximum target cement content
  (v) Type of cementitious material: special cement, ggbs, pfa
 (vi) Total air content of an air-entrained mix
(vii) Admixtures.

in addition to essential precautions such as concrete cover and curing.

The assumptions being made by the specifier are that, by specifying boundary values, usually selected from BS codes of practice, e.g. BS 8110 for structural concrete for buildings, appropriate to the severity of the exposure, then the chemical resistance and pore structure of the concrete in the actual construction will be adequately resistant for a chosen life of the structure. Naturally considerable imprecision is involved, and where the specifier fails to supervise the construction, optimism may be misplaced.

In respect of the quality of concrete provided to the construction, the absence of adequate tests which could be used by the specifier to confirm that the specified values were being met led BRMCA, BACMI, CMF and BCA to develop equivalent strength grades which could be used to reasonably guarantee minimum cement contents and water/cement ratios anywhere in the United Kingdom. Equivalent grades are detailed and discussed further in section 4.6.

Table 3.2 Summary of main BS 5328 recommendations in respect of chloride content

Type or use of concrete		Maximum total chloride content % by weight of cement
Reinforced (or containing embedded metal)	(i) Prestressed or heat cured concrete	0.1
	(ii) Sulphate-resisting Portland cement (BS 4027) concrete	0.2
	(iii) Other concrete	0.4
Other		—

*Inclusive of ggbs or pfa

Air-entrainment to provide resistance to frost attack and reduced permeability is considered under admixtures (section 1.3). Sulphate resistance is considered under cementitious materials (section 1.2).

As an alternative to a designed mix, the specifier may prescribe a mix for durability as follows.

(i) Target cement content
(ii) Type of cementitious material: special cement; ggbs; pfa
(iii) Total air content of an air-entrained concrete
(iv) Admixture.

3.4.4 *Corrosion of reinforcement*

To limit the risk of corrosion of reinforcement, BS 5328 has provided recommendations correlating degrees of exposure, grade, minimum cement content and maximum W/C coupled with limitations on chloride content. BS 8110 extends the correlation to include cover to the reinforcement.

3.4.5 *Chlorides in concrete*

BS 5328 bans calcium chloride in concrete to contain embedded metal and recommends the limits in Table 3.2 for chloride content from other sources.

When these limits are invoked by specifiers, control of chloride content of aggregates will usually be necessary and, in the case of the very low limits for prestressed concrete, it may mean that marine aggregates cannot be used without very special precautions. It will also be necessary to account for chloride in cement, admixtures and water.

3.4.6 *Alkali–silica reaction*

In many satisfactory concretes, some reaction can be detected between alkalis, mainly from cements, and reactive forms of silica in the aggregates. Only when the reaction, which is expansive, causes significant stresses and cracking of the concrete is it necessary to consider alkali–silica reaction (ASR) as deleterious. Fortunately, in the UK, serious damage due to this reaction has so far been reported only in the southwest and Midlands.

While specifiers in areas which are apparently safe may not consider the necessity for any special precautions, those in affected areas and those dealing with major structures where the conditions are likely to approach those known to have the potential for producing damage, may consider that some precautions should be taken.

The three conditions causing damaging ASR are moist condition of the concrete, sufficient alkali, and reactive silica.

All three need to occur before serious ASR will be manifested by disruption of the concrete. The alkalis originate primarily in the cement, but there is the possibility of significant contributions from other sources such as ggbs, pfa, admixtures or sea water.

The problem is considered under aggregates (section 1.1) and cement (section 1.2), further detailed advice is provided in the Concrete Society Report [22] and BRE digest 330 [145].

The alternative means of guarding against damaging ASR are as follows:

(i) Ensure concrete remains dry in service
(ii) Use a source of Portland cement of declared alkali level of 0.6% or below*
(iii) Use a sulfate-resisting Portland cement of guaranteed alkali content below 0.6%*
(iv) Use a combination of Portland cement and ggbs or pfa, with a total reactive alkali content of the cementitious material of 0.6% or less*
(v) Use at least 50% ggbs in the cementitious material, either added at the mixer or as a composite cement where the acid soluble alkali content of the cement or combination does not exceed 1.1% Na_2O equivalent
(vi) Use a combination of Portland cement and at least 25% ggbs or pfa *and* ensure that the reactive alkali content of the concrete is $3 \, kg/m^3$ or less
(vii) Ensure the cement and ggbs or pfa content does not exceed a value calculated from the reactive alkali level of the materials to maintain the alkali level of the concrete at $3 \, kg/m^3$ or below
(viii) Use aggregate and cement combinations with a known satisfactory history
(ix) Use aggregate combinations known to be unreactive (see aggregates section).

Experts are not in agreement over the most appropriate method for calculating the reactive alkali content for ggbs or pfa. Alternative methods based on water solubility and acid solubility are provided in BS 5328, and it is essential to check the basis of limits given in specifications.

---

*Provided that active alkalis from sources other than the cement do not exceed 0.2% by mass of cement.

4 Mix design

4.1 Principles of mix design

The general purpose of concrete mix design for a specific contract and use is to select materials and their proportions which will meet, economically, the properties of the concrete in both the fresh and hardened states.

For ready-mixed concrete in general, the purpose is extended, refined and divided into stages [80, 87].

(i) *Materials selection*
    To *select materials* which will meet most economically the requirements of most specifiers and intended uses.
(ii) *Mix proportioning*
    To *select proportions*, in anticipation, to cover the range of all likely specifications and uses
    To *determine the total quantities* of materials to ensure the ordered concrete volume is provided
    To *produce batch data* for the production of the full range of possible mixes and volumes which might be ordered
    To *produce mix data* to ensure designs are instantly available.
(iii) *Mix selection*
    To *select the most appropriate mix* to meet the specified requirement and intended use, taking account of the latest feedback from the quality control systems
    To provide additional data for *special mix designs* not covered by the above.

This total approach is much more comprehensive and sophisticated than that of the classic Road Note No. 4 or the present DoE Design of Normal Concrete Mixes [88], for which the main use is to allow a quick first assessment of a possible solution, with only limited availability of information on the materials and their performance.

Ready-mixed concrete is treated as a multi-component material, and the more sophisticated approach enables optimum proportioning to be achieved, resulting in the most economic composite solution for any given specification and use.

4.2 BRMCA method of concrete mix design

The BRMCA method [80, 87] has been developed over many years, has been adopted throughout the industry and has been used in the production and

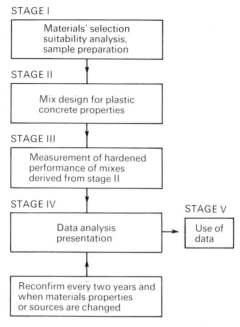

STAGE I

| Materials' selection
suitability analysis,
sample preparation |

STAGE II

| Mix design for plastic
concrete properties |

STAGE III

| Measurement of hardened
performance of mixes
derived from stage II |

STAGE IV   STAGE V

| Data analysis
presentation | → | Use of
data |

| Reconfirm every two years and
when materials properties
or sources are changed |

Figure 4.1   Major stages of the BRMCA method of mix design. After Dewar [80]

control of many hundreds of millions of cubic metres of concrete. The method is in widespread use by members of the Quality Scheme for Ready Mixed Concrete, and ensures that valid technical data are available to concrete users to substantiate the properties and quantities of concrete mixes being supplied from QSRMC Depots. The main stages in the BRMCA method are illustrated by a simple flowchart (Fig. 4.1).

A key feature of the BRMCA method is that in Stage II (Fig. 4.2), mixes are designed for optimum performance in the plastic state to ensure that the concrete is suitable for transporting, handling, compacting and finishing. As a result, a percentage of fine to total aggregate is decided upon for each mix which will provide sufficient cohesion, with a margin of safety, to minimize the risk of serious segregation under normal handling of the concrete. Only when the technologist is satisfied with the designs are the hardened properties of the mixes determined in Stage III for the laboratory method (Fig. 4.3), or from tests in production. In the alternative production batch method for determining strength, concrete is sampled from a substantial number of production batches over a wide range of cement contents during a period of 12 months.

The analysis and recording of trial mix and strength test data in Stage IV (Fig. 4.4) are based on cement content as the primary mix parameter to which all other mix constituents and concrete properties are directly related. The key relationships are illustrated, for typical materials, in Fig. 4.5, *a–f*.

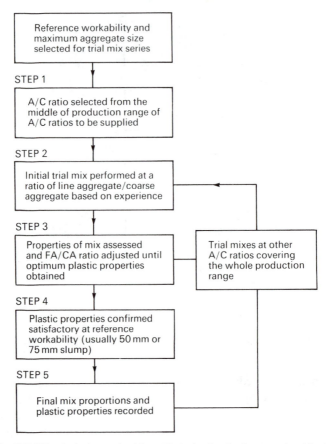

Figure 4.2   BRMCA mix design method Stage II–design for plastic properties. After Dewar [80]

The key relationships derived in Fig. 4.4 are those between (i) cement content and the major specified concrete properties (Fig. 4.5 *a–d, f*) and (ii) cement content and the weight of each constituent in the mix (Fig. 4.5*e*).

4.2.1 *Use of base data from BRMCA mix design method*

For use in concrete production, the base data shown graphically in Fig. 4.5*e* are stored in tables or graphs and in batch books or computer memory, in increments of cement content of either 5 or 10 kg/m³ for the typical production range of mixes from 100 to 450 kg/m³. The base data provide the ideal proportioning of mix constituents to achieve specified concrete properties, the correct batching instruction for the production of mixes, and provide the basis for altering mix proportions to achieve different values of workabilities and special concrete properties. They enable the key parameters for every

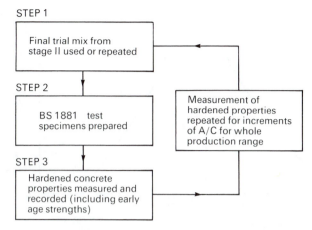

STEP 1

Final trial mix from
stage II used or repeated

STEP 2

BS 1881 test
specimens prepared

STEP 3

Hardened concrete
properties measured and
recorded (including early
age strengths)

Measurement of
hardened properties
repeated for increments
of A/C for whole
production range

Figure 4.3 BRMCA mix design method Stage III–performance of hardened concrete. After Dewar [80]

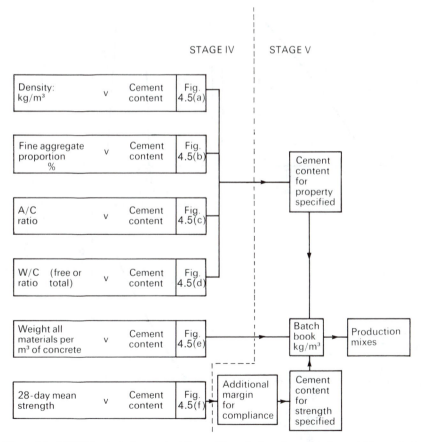

STAGE IV   STAGE V

Density:
kg/m³   v   Cement content   Fig. 4.5(a)

Fine aggregate proportion %   v   Cement content   Fig. 4.5(b)

A/C ratio   v   Cement content   Fig. 4.5(c)

W/C (free or total) ratio   v   Cement content   Fig. 4.5(d)

Weight all materials per m³ of concrete   v   Cement content   Fig. 4.5(e)

28-day mean strength   v   Cement content   Fig. 4.5(f)

Cement content for property specified

Batch book kg/m³

Production mixes

Additional margin for compliance

Cement content for strength specified

Figure 4.4 BRMCA mix design method Stages IV, analysis and presentation and V, use of mix design data. After Dewar [80]

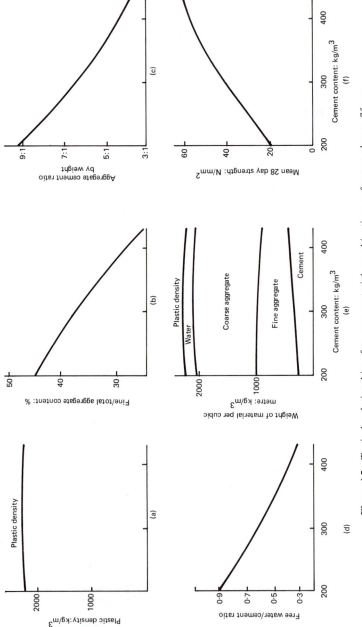

Figure 4.5 Typical relationships for one material combination: reference slump 75 mm, maximum aggregate size 20 mm. After Dewar [80]

combination of materials to be estimated, e.g. water/cement ratio or plastic density.

The relationship between cement content and strength allows the selection of a mix to satisfy any specified strength requirement when account is taken of the following aspects, amplified in Chapter 6 on quality control:

(i) In selecting the cement content of a designed mix, a target mean strength has to be selected that exceeds the specified characteristic strength by a design margin which includes a statistical constant and the depot standard deviation, in the following way:

$$\begin{array}{ll} \text{target mean} = \text{specified} + \text{design} \\ \text{strength} \quad\quad \text{strength} \quad \text{margin} \end{array}$$

where design margin = depot standard deviation × k

(ii) The value of the statistical constant k is selected according to the proportion of test results that are permitted to fall below the specified strength and to accord with the specified compliance rules.

(iii) Typically, a value for k of 2 is adopted by the producer to satisfy compliance rules of BS 5328 or BS 8110 (CP 110 revised), but on occasion a higher value may be selected to build in additional safety. Incidentally, the principle of characteristic strength, adopted by BS 8110, requires k to be only 1.64.

(iv) In all cases where designed mixes are supplied, the producer has the responsibility of controlling the quality of the concrete so that the compliance requirements are satisfied. To achieve this objective, the producer operates a QSRMC-approved control system which compares the results of tests on samples of materials and concrete with the performance of the materials and concrete on which the mix design was based.

(v) The producer adjusts the proportions of the mix to correct for any significant deviation between these measurements.

(vi) The concrete control system also provides a current value of standard deviation for use in the formula.

(vii) The cement content is adjusted to take account of the workability specified relative to that adopted, usually 50 or 75 mm, in establishing the relationship.

The continuous measurement and analysis of data in the control system to ensure compliance with specifications and British Standards also ensure that the relationships derived in the trial mix programme are continuously confirmed and the need for any modification swiftly identified.

The use of base data from the BRMCA method of mix design for determining the batch weights per cubic metre of concrete for a designed mix specification is illustrated by the example in section 4.2.2 following.

4.2.2 Example: selecting batch proportions for specified design mix requirements [80, 87]

Select batch quantities for a designed mix to BS 5328 for a typical mix relationship derived using the BRMCA method. Specified strength grade C30, maximum free water/cement ratio 0.50 and slump 50 mm (see Fig. 4.6).

Step 1: Select target mean strength TMS
TMS = 30 + (SD × k)
k = 2 for BS 5328 compliance rules
Use SD = 5.0 N/mm² from quality control information for standard deviation
TMS = 30 + (5.0 × 2) =   40 N/mm²
Step 2: Read off cement content for TMS of 40 N/mm².   300 kg/m³
Step 3: Read off cement content for free W/C ratio of 0.50   360 kg/m³
Step 4: Use highest cement content from step 2 or step 3   360 kg/m³
Step 5: Read off mix quantities per cubic metre
   for cement content of 360 kg/m³ cement   360 kg/m³
Fine aggregate*   620 kg/m³
Coarse aggregate*   1200 kg/m³
Free water   180 kg/m³
*Saturated and surface dry

Note: In this typical example, a cement content of 360 kg/m³ is required to achieve the specified free W/C ratio of 0.50 which will achieve a mean 28-day strength of about 50 N/mm² (Step 6).

4.2.2.1 Benefit of specifying a higher grade. For the particular materials in the example, specification of a BS 5328 strength grade C40 (50 N/mm² less margin

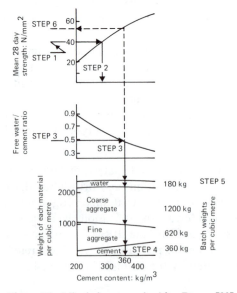

Figure 4.6   Mix design example. After Dewar [80]

of 10 N/mm²) instead of C30 would enable the free water/cement ratio of 0.50 to be assured by testing the concrete for strength compliance in accordance with BS 5328.

4.2.2.2 Adjustments to batchweights to allow for moisture content of aggregates.
Because aggregates are almost invariably batched at a higher moisture content than the saturated and surface dry (ssd) condition, it will be necessary to increase the batched weight of the fine and coarse aggregates by their respective free moisture contents, to ensure the correct ssd weights. The amount of water needed to be added by the batcherman will be reduced correspondingly. Using the designed mix example and assuming 7% free moisture in the fine aggregate and 3% free moisture in the coarse aggregate, the adjustments to the batch figures would be as follows:

Material	Mix data (kg/m³)	Moisture (kg)	Adjusted batch data
Cement	360	—	360
Fine aggregate	620 (ssd)	$\frac{7}{100} \times 620 = 43$ (say 40)	670 (wet)
Coarse aggregate	1200 (ssd)	$\frac{3}{100} \times 1200 = 36$ (say 40)	1240 (wet)
Water	180 (free)	80	100 (added)
Total	2360		2360

4.2.3 *Selection of batch proportions for a prescribed mix*

Batch proportions for a prescribed mix are made using a process similar to that in the following example for a BS 5328 standard mix.

4.2.4 *Example: selecting batch proportions for a standard mix*

Select batch quantities for a standard mix to BS 5328 for a typical mix relationship derived using the BRMCA Method. ST4 mix 20 mm aggregate, 75 mm slump (see Table 4.1 and Fig. 4.7).

Step 1: Select cement content from Table 4.1 for ST4 mix   300
Step 2: Read off mix quantities for cement content of 300 kg/m³
         cement   300 kg/m³
         fine aggregate (ssd)   700 kg/m³
         coarse aggregate (ssd)   1160 kg/m³
         water (free)   180 kg/m³

Note: The cement-contents of BS 5328 standard mixes are determined from Table 1 of BS 5328 (see Table 4.1). The actual performance of the mix is determined by the materials used.

The base data have established the ideal ratio of fine aggregate to coarse aggregate to achieve the required plastic concrete properties and yield. The total quantity of aggregate is 1870 kg, close to the guidance value of 1860 kg provided in BS 5328 for the ST4 mix.

As for the designed mix example, adjustments to batch weights would be necessary to allow for the moisture contents of wet aggregates.

4.2.5 *The predictable future*

The BRMCA method of mix design requires considerable expenditure of effort in the laboratory for the making of trial mixes both initially and periodically thereafter to confirm or modify the initial data. A computerized method [12] has been developed for use by BRMCA members which enables them to predict the results for properties of fresh concrete on the basis of materials properties alone. Co-operation by manufacturers of cement, aggregates, ggbs and pfa in providing data enabled this method to be introduced during 1984.

In this method, the properties to be determined for each material are:

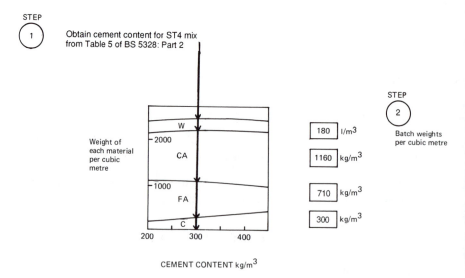

Figure 4.7 Mix design example for a standard mix to BS 5328.

Table 4.1 Extract from BS 5328 for standard mixes

Standard mix	Nominal maximum size of aggregate slump (mm)	40	20
	Specified slump (mm)	75	75
ST1	Cement (kg)	180	210
	Total aggregate (kg)	2010	1940
ST2	Cement (kg)	210	240
	Total aggregate (kg)	1980	1920
ST3	Cement (kg)	240	270
	Total aggregate (kg)	1950	1890
ST4	Cement (kg)	280	300
	Total aggregate (kg)	1920	1860
ST5	Cement (kg)	320	340
	Total aggregate (kg)	1890	1830

Note: The values for total aggregate are for guidance only

Mean size	(determined from the grading)
Voids ratio	(determined for aggregates from tests for bulk density and relative density; determined for powders from standard consistence tests of pastes)
Relative density	

The values for each material are processed by computer by use of a complex mathematical model which assesses the effects of combining them together in concrete. This leads to a computer output of optimized mix proportions over the full range of possible cement contents and water/cement ratios. The optimization ensures correct proportioning for adequate cohesion and correct yield.

An example of the accuracy of prediction is shown in Table 4.2 where actual laboratory trial mix data are compared with predictions from materials properties. From such predictions, computerized batch book data covering the full range of cement content and batch size can be prepared. The method can be used for confirming existing batch data, for updating existing batch data or for producing new data for new materials and plants.

4.2.6 *Technical advice on concrete properties based on the BRMCA method*

On request, and as required by BS 5328, purchasers of ready-mixed concrete may be supplied by the producer with data derived from the mix design process. In addition, the concrete producer is able to advise on a wide range of concrete properties and uses by developments from the base data, for example:

Uses of concrete	Pumping
	Flowing concrete
	Special handling,
	placing and compaction
Other materials	Aggregate size
and performance	Special aggregates
	Admixtures
	ggbs or pfa.

4.3 Mix design using ggbs or pfa

Cement, fine and coarse aggregates provide a three-component solids system enabling void content, and thus water demand, to be minimized by appropriate selection of mix proportions of the three solid components. This is an essential part of the skill of the ready-mixed concrete technologist who is responsible often for tens of thousands of cubic metres of optimized concrete annually.

The advent of other cementitious materials such as ggbs or pfa, far from being a nuisance, provide an opportunity for the mix designer to show his skill further with a four-component solids system, enabling further reduction in voidage and water demand.

For ggbs or pfa concretes, either a similar comprehensive mix design approach is adopted, or well-tried adjustments are made to accurate plain concrete mix quantities to allow for the properties and proportions of ggbs or pfa to be used, as in the example below for pfa mix design.

4.3.1 *Example: an adjustment to a mix design for the use of pfa* [38]

It is required to modify the quantities per cubic metre of a plain concrete, known to meet the specified strength, to include 30% of pfa by weight of cementitious material and to have the same slump and 28-day strength. It is assumed that the following data have been accumulated for a particular quality of pfa to BS 3892: Part 1 and for this level of pfa inclusion.

$$\text{Increase in cementitious material} = 12\%$$
$$\text{Water content reduction} = 5\%$$
$$\text{Relative density of cement} = 3.12$$
$$\text{Relative density of pfa} = 2.25$$
$$\text{Relative density (SSD) of fine aggregate} = 2.59$$
$$\text{Relative density (SSD) of coarse aggregate} = 2.50$$
$$\text{Standard deviation is assumed unaffected}$$

Table 4.2 Comparison of actual concrete test data with data estimated by computerized simulation based on current materials tests [80]

Parameter	Fines: %				Free water: $1/m^3$				Density: kg/m^3			
Cement: kg/m^3	140	250	340	460	140	250	340	460	140	250	340	460
Concrete test data	39	37	33	26	180	157	154	172	2295	2343	2372	2396
Computerized simulation	40	36	31	21	182	160	161	178	2303	2357	2371	2364
Computerized simulation minus concrete test data	+1	−1	−2	−5	+2 (+1%)	+3 (+2%)	+7 (+4%)	+6 (+3%)	+8 (0%)	+14 (+1%)	−1 (0%)	−32 (−1%)

Plain concrete

OP cement	350 kg/m³
Fine aggregate (SSD)	625
Coarse aggregate (SSD)	1160
Water (free)	180

$$2315 \text{ kg/m}^3 \text{ (air 1\%)}$$

Pfa concrete

	Wt. (kg/m³)	Vol. (m³)
Total cementitious material	$= 1.12 \times 350 = 392$	
OP cement	$= 392 \times 0.70 = 274$	$\dfrac{274}{3120} = 0.088$
pfa	$= 392 \times 0.30 = 118$	$\dfrac{118}{2250} = 0.052$
Water (free)	$= 180 \times 0.95 = 171$	$\dfrac{171}{1000} = 0.171$
Air	$=$	0.010
Total	$=$	(0.321)
Aggregate	$=$	$1 - 0.321 = (0.679)$
Coarse aggregate (SSD)	$=$ (unaltered) 1160	$\dfrac{1160}{2500} = 0.464$
Fine aggregate (SSD)	$=$	0.679 $- 0.464 = 0.215$
	$= 0.215 \times 2590 = 557$	
Total	$=$ 2280	1.000

Comparison (kg/m³)

	Plain concrete	pfa concrete
OP cement	350	274
pfa	—	118
Fine aggregate (SSD)	625	557
Coarse aggregate (SSD)	1160	1160
Water (free)	180	171
	2315	2280

It may be noticed that, for the pfa concrete, the total cementitious material is greater but the OP cement content is smaller, the fine aggregate content is reduced, the coarse aggregate content is deliberately the same, the water is reduced and the density is reduced because of the lower density of pfa compared with cement.

Note that it should not be assumed that ggbs concretes or other pfa concretes would require the same adjustments. The factors can differ appreciably between materials, sources and qualities and will be influenced by the proportion of ggbs or pfa, the cement content and other factors. The method, however, will be applicable and can be used for any situation, provided the factors are known.

4.4 Judging concrete mix design

A major difficulty exists for the inexperienced specifier in finding a standard against which to judge a proposed mix design. He may, quite naturally, reach for a copy of the excellent 'Design of normal concrete mixes' [88] and be quite dismayed to discover that the estimated mix proportions bear no resemblance to those being proposed. The usual reasons are that, for the initial design, 'Design of normal concrete mixes' assumes:

(i) Constant water content over the complete range of cement content for constant slump

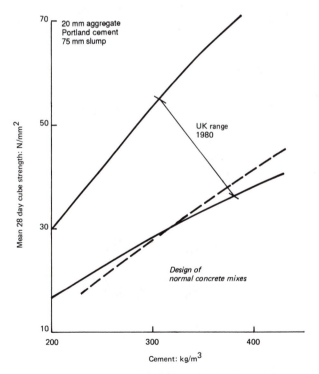

Figure 4.8 Comparison of assumed relationship from 'Design of normal concrete mixes' and range of actual relationships in 1980 for ready-mixed concrete. After Dewar [80]

(ii) Relatively low cement strengths compared with typical values available today.

The first assumption was made for simplicity, but can lead to appreciable error. The second generally leads to appreciable over-estimation of cement content. This was quite deliberate, ensuring safe designs for use with almost any materials anywhere in the UK, and has been further affected by increased cement strength since publication of the booklet. When the initial design is followed by laboratory mixes, as recommended by the booklet, appreciable adjustments will normally be found to be required.

Specifiers may find it useful to compare proposed mix designs with the envelope of data in Fig. 4.8 obtained from a survey made by BRMCA in 1980 for 75 mm slump concrete. Also superimposed on this diagram is a typical relationship from 'Design of normal concrete mixes'. The distribution of relationships from the survey was not symmetrical, so that a greater proportion occurred in the upper part of the band. Thus, typically, there is a greater probability of the specifier finding a present-day mix design in this upper part. However, in some areas of the UK it is possible for the combination of cement and aggregate qualities to yield a relationship towards the lower boundary.

Note that Fig. 4.8 relates achieved *mean* strength to cement content. Commonly, target mean strength is set about $10 \, \text{N/mm}^2$ higher than the specified strength.

4.5 Water/cement ratio—free or total

Water/cement ratio can be measured either in terms of free or total water, the difference being the water absorbed by the aggregate. The difference can be significant, so that it is important for the specifier to make clear the basis (free or total) and to query the basis of mix design data when it is not made clear which is being quoted by the producer.

$$\text{free water} = \text{total water} - \text{absorbed water}$$

$$= \text{total water} - \text{aggregate content} \times \frac{\text{absorption}(\%)}{100}$$

$$\text{free W/C} = \text{total W/C} - (\text{A/C}) \times \frac{\text{absorption }(\%)}{100}$$

So, for example, if total W/C is 0.56, A/C is 6 and percentage absorption for the combined fine and coarse aggregate is 1%, then

$$\text{free W/C} = 0.56 - 6 \times \frac{1}{100} = 0.50$$

'Free W/C' is usually assumed to be the effective water/cement ratio in concrete determining the particle spacing of cement in concrete and influencing porosity, permeability, durability and strength of the concrete.

Unfortunately, it is difficult to measure 'free W/C' accurately, and virtually impossible to measure it directly. Reliance has to be placed upon indirect measurement via total water content and aggregate absorption, which is dependent upon the subjective assessment of the saturated and surface-dry (SSD) condition of the aggregate.

4.6 Meeting durability requirements

Generally, specification requirements for durability are written in terms of minimum values for cement content and maximum values for free water/cement ratio. Values for these parameters estimated to meet strength and workability requirements are checked by the producer against the specified limits before making final mix design proposals.

Very often limits set for cement content and water/cement ratio entail achieving a much higher strength than the grade specified, so that the specifier cannot use observed compliance with specified strength as a means of assuring himself that the limits set for durability have been met.

4.6.1 Equivalent strength grades—ensuring durability

The absence of simple and accurate tests which can be used by the purchaser to check whether the concrete complies with a specified minimum cement content or maximum water/cement ratio has been a cause of concern to the responsible ready-mixed concrete producer.

To overcome this problem, BRMCA, in co-operation with BACMI, the Cement Makers Federation and the BCA, developed the concept of national or equivalent grades by which strength grades can be specified which have better than 95% probability of ensuring compliance in the UK with any

Table 4.3 Equivalent grades for 10 to 40 mm aggregate, cement of 42.5 strength class to BS 12 and 500 to 150 mm slump

Minimum cement (kg/m^3)	Maximum free water/ cement ratio	Equivalent grade	Approximate mean strength* N/mm^2
200, 210	—	C15	20–25
220, 230	—	C20	30
240, 250, 260	0.70	C25	35
270, 280	0.65	C30	40
290, 300, 310	0.60	C35	45
320, 330	0.55	C40	50
340, 350, 360	0.50	C45	55
370, 380, 390	0.45	C50	60

*For guidance only

specified cement content or water/cement ratio [81, 89]. Examples of such nationally applicable grades included in BS 5328 are shown in Table 4.3.

By selecting the highest appropriate grade for his needs, the specifier may use cube tests to confirm that the cement content and free water/cement ratio have a very high probability of being in compliance anywhere in the UK. Some approximate indications of the mean strengths corresponding to the various grades are also provided in Table 4.3, to aid judgement. BS 8110 has adopted grades compatible with this table.

5 Statistics for quality control, mix design and compliance

A knowledge of certain aspects of statistics is essential for understanding and operating quality control and mix design of ready-mixed concrete. A far greater knowledge is required for assessing the implications of compliance clauses in specifications. In this section, an introduction is provided to help those involved in quality control and mix design, and to show some relevant results of using more complex concepts in assessing BS 5328 compliance clauses.

5.1 Statistical terms

Various statistical terms and measurements, developed by mathematicians and statistical practitioners, are common to the assessment and control of variation of many measured parameters for a wide range of natural and manufactured materials or products.

The terms most commonly encountered are:

Variation
Repeatability and reproducibility
Precision and tolerance
Distribution
Normal distribution
Histogram
Mean
Range
Standard deviation
Coefficient of variation
Operating characteristic

5.2 Variation

If one technician prepared a number of representative samples of concrete from one batch of concrete and measured the slump twice on each sample, it would be very unlikely that each of any pair of slump values would be identical or that all the mean values of the pairs from the different samples would be the same. If the same technician repeated the test but used a different standard slump cone or a second technician made tests at the same time, it would again be unlikely that identical results would be obtained. In answer to the question: 'What is the slump of the batch of concrete?' each technician can say only 'My estimate of slump is...'.

If the first technician now goes on to test many different batches of concrete, all intended to be identical, different slump values will again be obtained. The differences will be due now not only to the inherent variation of the test but also due to real variations in slump between the different batches. Tolerances about specified slumps are provided to take account of both types of variation.

Similarly, if the technician makes air tests or casts cubes, the results will show variation in air content or strength, some of the variation being due to the method of sampling and testing and some being due to variation in properties between different batches of concrete.

British standards for concrete testing increasingly provide data on repeatability and reproducibility, which are both aspects of precision. The first term relates to variation associated with the individual and the equipment used, the second to variation *between* individuals, laboratories and equipment. Data on precision enable tolerances to be determined for inclusion in specifications.

The precision of a result cannot be assessed from a single test, which is why British Standards rarely permit a single test result to be used for compliance purposes. Because of the imprecision of some tests, British Standards may insist on obtaining a number of results, only the average of which may be used to assess compliance. Even so, tolerances often need to be quite wide to account for imprecision.

5.3 Distribution

If results of cube strength are compared for different batches of the same mix design, they may or may not be seen to be grouped in any particular way, but as the number of results increases, a pattern or distribution will usually emerge of the type shown in Fig. 5.1. The block diagram, called a *histogram*, can be seen to peak near the middle, fewer results occurring as one moves away from the middle. The diagram is sensibly symmetrical: whatever happens on the left, also happens on the right.

5.4 Normal distribution

The symmetry and shape of Fig. 5.1 are completely normal and would be expected to occur for many other measurements on materials, not only with concrete. In fact, they are so normal that mathematicians have discovered mathematical laws by which the shape can be drawn on the basis of just one measurement, called standard deviation, the calculation of which is shown later. The standard deviation is usually about 1/4 to 1/6th of the spread of the complete histogram, provided there are plenty of results in it. For the

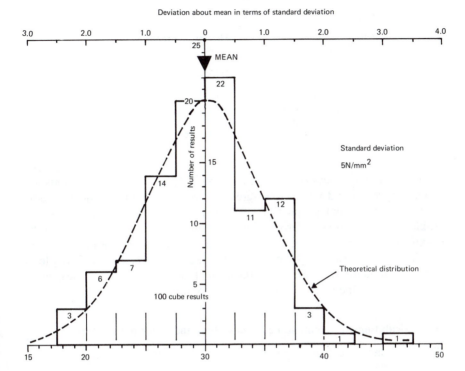

Figure 5.1 Comparison of the theoretical distribution and a histogram for 100 cube results when standard deviation is 5 N/mm² and the mean is 30 N/mm². Note: The histogram has been obtained by counting the numbers of results in columns 2.5 N/mm² wide. The numbers are indicated at the top of each column.

histogram in Fig. 5.1, the standard deviation can be shown to be 5 N/mm², which accords with the spread observed of about 30 N/mm².

Mathematical laws predict that the proportion of results lying within various limits on either side of the middle will be as in Table 5.1. Because

Table 5.1 Proportion of results expected to lie between pairs of limits.

limits (about the mean)	Theoretical proportions between limits	Actual proportions between limits in Fig. 5.1
± 0.5 × standard deviation	38%	42%
± 1 × standard deviation	68%	67%
± 1.5 × standard deviation	86%	86%
± 2 × standard deviation	95%	95%
± 2.5 × standard deviation	99%	99%
± 3 × standard deviation	100% approx	99%

Table 5.2 Proportions of results expected to lie below a lower limit.

Limits below the mean	Theoretical proportions below limits
0.5 × standard deviation	31%
1 × standard deviation	16%
1.5 × standard deviation	7%
2 × standard deviation	2.5%
2.5 × standard deviation	0.5%
3 × standard deviation	nil approx.

concrete strength obeys the same laws applicable to other measurements, we would expect to find reasonable agreement between the theoretical proportions and the actual proportion of results, as is shown to be the case in Table 5.1. The theoretical proportions of results below any limit less than the mean value can be obtained from Table 5.2. It will be seen from Fig. 5.1 that 3% of results are below 20 N/mm² close to the 2.5% expected from theory for a limit 2 × standard deviation below the mean. Note that in this case the mean is 30 N/mm² and the standard deviation is 5 N/mm².

5.5 Calculations of mean, standard deviation and other parameters

The formulae for calculating mean and standard deviation are:

$$\text{Mean } (\bar{x}) = \frac{\text{Sum of measurements}}{\text{Number of measurements}} = \frac{(\Sigma x)}{n}$$

$$\text{Standard deviation} = \sqrt{\frac{\Sigma(x \sim \bar{x})^2}{(n-1)}}$$

There are derived formulae for use in calculators or computers which will give the same answers. There are also graphical methods [90]. The expression in the numerator of the formula for standard deviation is shorthand for sum (Σ) of squares ( )² of differences ($\sim$) between each result (x) and the mean ($\bar{x}$). The 'minus one' in the denominator of the formula is a correction to produce the best estimate of standard deviation, to allow for the fact that the true value of the mean is not known, ($\bar{x}$) being only an estimate.

Calculators, and of course computers, can be used to calculate mean and standard deviation quickly with minimum risk of error, provided the correct data are fed in. It is necessary to check that the formula used by the calculator or computer contains $(n-1)$ and not (n) in the denominator. Use of the $(n-1)$ formula will be necessary when small numbers of results are involved, but as n increases in size, so the error is reduced and might safely be ignored. An accurate assessment of standard deviation requires a substantial number of results, never less than 25 and preferably 100 or more.

To understand how the formula operates, an example is provided in

Table 5.3 Calculation of standard deviation

28-day strength (N/mm^2) (x)	Results above mean (x − x̄)	Results below mean (x̄ − x)	(x ∼ x̄)^2
40.5	5.5		30.25
36	1		1
33		2	4
38	3		9
42	7		49
32.5		2.5	6.25
29		6	36
37	2		4
28		7	49
36	1		1
Σ(x) = 352	Σ(x − x̄) = 19.5	Σ(x̄ − x) = 17.5	Σ(x ∼ x̄)^2
n = 10			= 189.5
x̄ = 35.2 (say 35)			

Table 5.3 for only 10 cube results. The method is the same for 100, or indeed any number of results.

Note that the two centre columns should add to approximately the same number. This is useful as a check on calculation up to that point. Any difference should be due only to approximation of x̄.

$$\text{Mean } \bar{x} = \frac{\Sigma x}{n} = \frac{352}{10} = 35.2, \text{ say } 35 \text{ N/mm}^2$$

$$\text{Standard deviation (s.d.)} = \sqrt{\frac{\Sigma(x \sim \bar{x})^2}{n-1}} = \sqrt{\frac{189.5}{9}} = \sqrt{21.06} = 4.59$$

$$= \text{say } 4.5 \text{ (to nearest} \sim 0.5 \text{ N/mm}^2)$$

$$\text{Coefficient of variation} = \frac{\text{s.d.}}{\bar{x}} \times 100\% = \frac{4.5}{35} \times 100\% = 12.8, \text{ say } 13\%$$

$$\text{Range} = \text{highest } (x) - \text{lowest } (x) = 42 - 28 = 14 \text{ N/mm}^2$$

5.6 Sources of variation

A typical breakdown of standard deviation into its component parts is illustrated by Table 5.4. Thus approximately similar importance attaches to the four components with an emphasis on production and cement. Major improvement in the overall s.d. requires major improvements in all four components.

Short-term standard deviations for 28-day strength of ready-mixed concrete range from 3.5 to 6.5 N/mm^2 with a typical value of, say, 4.5 N/mm^2 [91] or 5 N/mm^2.

Table 5.4 Typical components of standard deviation for concrete strength.

Aspect	Standard deviation (N/mm²)
Cement	2.5
Aggregates	2
Sampling and testing	2
Production	2.5
Overall	4.5

Note: Overall s.d. is square root of sum of squares of component values.

The value of 4.5 N/mm², reported in 1980, is significantly lower than an average contract value of about 5.7 N/mm² reported [92] in 1970 for ready-mixed concrete delivered to highway structures; at that time only about 5% of reported values were less than 4.5 N/mm², as shown in Fig. 5.2.

5.7 Influence of mean strength level on standard deviation

Overall standard deviation for strength tends to be constant for medium and high strengths and decreases at low mean strengths, as shown in Fig. 5.3. For a mean strength in the range zero to about 27 N/mm², standard deviation

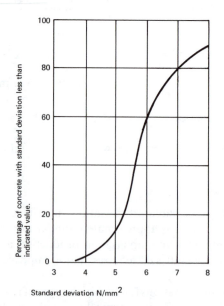

Figure 5.2 Distribution of standard deviation for ready-mixed concrete in highway structures reported in 1970. Redrawn and adapted from Metcalf [92]

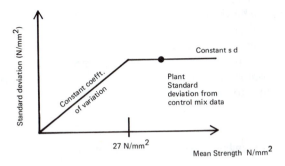

Figure 5.3 Assumed relationship between standard deviation and mean strength of concrete

increases proportionally with mean strength, which implies that coefficient of variation is constant in this range, whereas above $27\,N/mm^2$, standard deviation is constant [41]. Ready-mixed concrete depot standard deviation is normally measured on concrete of mean strength above $27\,N/mm^2$, and thus quoted standard deviation values can usually be assumed to represent most structural concrete production.

In the special case of very lean low-workability concretes, variation due to compaction can be very significant, so that it is unwise to assume that lower standard deviations are applicable without confirmation directly from relevant test data.

5.8 Standard deviation due to sampling and testing

Standard deviation due to sampling and testing alone seems to vary proportionally with mean strength over the full range of strength, and thus is more significant for high-strength concrete.

The coefficient of variation for careful laboratory sampling and testing of cubes is approximately 3%. Thus standard deviation is about $1\,N/mm^2$ at a mean strength of $30\,N/mm^2$, and $2\,N/mm^2$ at $60\,N/mm^2$ under ideal conditions. Appreciably higher values can occur with poor site sampling and testing.

Table 5.5 Influence of standard deviation due to sampling and testing, on the observed overall standard deviation for strength of ready-mixed concrete.

Standard deviation due to sampling and testing (N/mm²)	Ready-mixed concrete observed overall standard deviation (N/mm²)			
	Standard deviation of concrete (N/mm²)			
	2	3	4	5
1	2.2	3.2	4.1	5.1
2	2.8	3.6	4.5	5.4
3	3.6	4.2	5.0	5.8

The importance for the producer of achieving a low standard deviation for sampling and testing in order not to inflate the observed overall standard deviation for ready-mixed concrete may be seen from Table 5.5. If the standard deviation is high for sampling and testing made on behalf of the purchaser for compliance, this can result in a misleading impression of a producer's ability to control uniformity. It may also lead to apparent compliance failures.

5.9 Relevance of standard deviation for concrete mix design

For designed mixes, it is necessary to estimate a target mean strength substantially in excess of the specified strength to reduce the risk of failing compliance clauses in specifications to an acceptably low level. Compliance clauses in BS 5328 are based on providing an acceptably low risk to the specifier of receiving more than 5% of concrete below the specified strength. Historic and recent experience have indicated that design for an average of 97.5% of concrete above the specified strength is an acceptable balance for the producer between economy and risk of failing the compliance clauses.

From Table 5.2 it will be apparent that, if the margin between target mean and specified strength is $2 \times$ standard deviation, the requirement of 97.5% of results above will be met. Thus, the well-tried formula is

$$\text{target mean strength} = \text{specified strength} + (2 \times \text{standard deviation})$$

where standard deviation is the plant standard deviation for target values above $27 \, \text{N/mm}^2$, and as follows for target values below $27 \, \text{N/mm}^2$ (see Fig. 5.3):

$$\text{s.d.} = \frac{C}{\dfrac{27}{\text{plant s.d.}} - 2}$$

where C is the specified strength.
For example, if $C = 15 \, \text{N/mm}^2$

$$\text{plant s.d.} = 4.5 \, \text{N/mm}^2$$

$$\text{s.d.} = \frac{15}{\dfrac{27}{4.5} - 2} = 3.75 \, \text{N/mm}^2.$$

5.10 Statistical implications of compliance rules

The purpose of a compliance rule is to provide a means by which the client can judge, with an acceptably low risk of error, whether complying concrete has been provided. It is normal today to do this in two ways: by providing one rule for every tested batch and a second rule covering a group of tested batches together with all those untested in between. Normally these are written as absolute or 'go–no go' rules rather than rules permitted to be failed a small proportion of the time. The unfortunate result is that, although the risk of any batch failing is a known constant, the probability of failure occurring at some time during a contract increases with the size of the contract. Naturally, the risks could be reduced by increasing the margin for larger contracts.

If the producer designs for 97.5% of concrete to be above the specified strength when the theoretical basis of the compliance rules is 95%, the risk of failing at all on any normal size of contract can be shown to be acceptably low. Thus, by this approach, the ready-mixed concrete industry runs less risk of problems with each individual small contract and also safeguards larger contracts and its overall production.

The conventional way by which probabilities may be judged by both producer and consumer is through the use of the type of diagram called an *operating characteristic curve* (Fig. 5.4). For any compliance clause, the producer may assess the risk of having complying concrete rejected (*producer's risk*) and the consumer can assess the risk of accepting non-complying concrete (*consumer's risk*) [93, 94].

In the theoretical diagram (Fig. 5.4a), the producer's risk and consumer's risk are both nil. In practice, diagrams are usually of the form of Fig. 5.4b, where both run some risk.

Note that the statistics associated with the construction of operating characteristic curves are not within the scope of this book—see [95], [101].

It is important not to treat as *failures* those batches of concrete represented by the 2.5% or 5% of results falling below the specified characteristic strength. They are permitted to occur. Indeed, their achievement is successful design, not failure. Only batches or sets of batches which fail compliance clauses are eligible to be treated as failures and their future subject to further consideration.

Note that 5% of results below the specified strength is an acceptable situation because this is actually the structural design criterion. Only the occurrence of more than 5% below constitutes a potential violation of structural criteria.

5.10.1 *Current compliance rules of BS 5328*

The current strength compliance rules of BS 5328 are

(i) No result for a batch shall be less than the specified strength less 3 N/mm^2
(ii) No mean of four consecutive results shall be less than the specified strength plus 3 N/mm^2

where *a result is the mean for a pair of cubes* from a single batch and tested at 28 days. BS 5328 also includes rules for the means of the first two and the first three results.

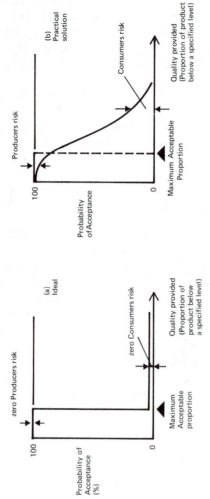

Figure 5.4 Ideal and practical operating characteristic curves for a compliance rule showing producer's risk (of having concrete rejected) and consumer's risk (of accepting unsatisfactory concrete)

Purists and even some practitioners will quite correctly point out that the value of 3 N/mm² used in each rule should really be some function of standard deviation. However, standard deviation can only be calculated accurately by the purchaser well into a large contract, and a value would have to be provided by the producer from his own data, which is tantamount to requiring him to make his own hangman's rope. The impracticality of this solution has resulted in the adoption by BSI of a constant addition or subtraction. Any slight advantage or disadvantage associated with high or low standard deviations is far outweighed by the simplicity of operating the rules in practice.

The producer/consumer risks may be seen for applications of each clause separately in Figs. 5.5 and 5.6. Account has been taken of the wording of the second clause, which permits it to be applied up to four times to each tested batch and up to three times to intermediate untested batches. Obviously, the greater the rate of compliance testing and of the application of a clause, and because of the 'go–no go' nature of the clause, the greater is the risk of concrete failing at some time during a contract. Also, under the 'means of four' clause, when testing is spread, the number of untested batches at risk under a single decision could have serious implications.

As a result, the future of a large quantity of concrete construction could be at risk, dependent upon the results of a few tested batches. The commercial consequences could be devastating as a result of a chance event, an error in production or inattention to correct sampling and testing by the purchaser.

BS 5328 attempts to cover this aspect by requiring the rate of testing to be declared by the specifier and also by limiting the maximum quantity of concrete at risk under a single application of the clause. When the same concrete is assessed by compliance clauses more than once, the producer's risk is substantially increased.

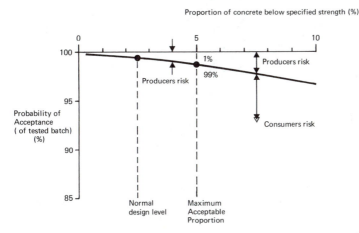

Figure 5.5 Producer's and consumer's risks for BS 5328 compliance rule (*a*) relating to individual results. Redrawn and adapted from Dewar [87]

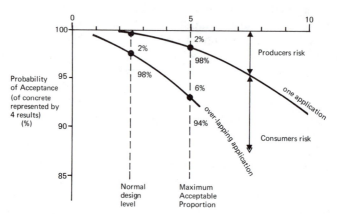

Figure 5.6 Producer's and consumer's risks for BS 5328 compliance rule (b) relating to the means of four results. Redrawn and adapted from Dewar [87]

It will be seen from Fig. 5.7 that there is a very low probability of a trouble-free contract when the contract is large unless the design proportion below the specified strength is reduced appreciably below the 5% permitted, and it may need to be below the usual 2.5%, unless a high risk is accepted by the producer.

Interesting situations can arise with multiple assessments over a period of

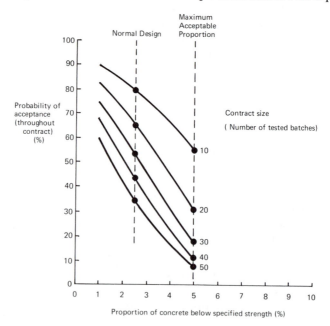

Figure 5.7 Probability of a trouble-free contract under application of BS 5328 compliance clause (b) relating to means of four results. Based on data from Dewar [87]

Table 5.6 Example of application of the BS 5328 clause (b), 'means of four', to successive cube results for grade C20 concrete

Cube no.	Results (N/mm²)	Mean of 4	Accept concrete represented by	Reject concrete represented by
1	34	—	—	—
2	31	—	—	—
3	25	—	—	—
4	27	29.2	1–4	—
5	20	25.8	2–5	—
6	25	24.2	3–6	—
7	21	23.2	4–7	—
8	23	22.2	—	5–8
9	30	24.8	6–9	—

Note: Criterion of acceptance is $20 + 3 = 23 \, \text{N/mm}^2$ for mean of four.

time. For example (see Table 5.6), in the case of four consecutive assessments, the same concrete can be judged to comply on three occasions and fail on the fourth, by which time further construction may have hidden it. The possible site scenarios are left to the reader's imagination and ingenuity.

5.10.2 *Influence on producer's risk of faults in sampling and testing for compliance*

The importance for the producer of standard sampling, specimen-making, storing and testing by the consumer or his representatives cannot be too strongly emphasized.

If non-standard techniques have been used which depress the results for strength by as little as $0.5 \times$ standard deviation, the proportion of results below the specified strength can change from an expected 2.5% to 7% and alter the probability of failing compliance with the BS 5328 'mean of four' clause (overlapping application) from 2% to 10% or more—see Fig. 5.6 and [95].

BS 5328 now requires that the range of the two cubes made from the same batch should not differ by more than 15% of the mean of the pair.

6 Quality control

The term 'quality control' means different things to different people. It is sometimes thought to be synonymous with the whole area of quality assurance or may merely be restricted to analyses of test data. The following definitions may clarify the meanings used here for ready-mixed concrete.

Quality Assurance: the operation of systems by which the client can be assured that control is being applied over all aspects of the complete process from enquiry to delivery to ensure compliance of the product with the specification agreed with the client. This involves the whole management team and their staffs together with independent inspection.

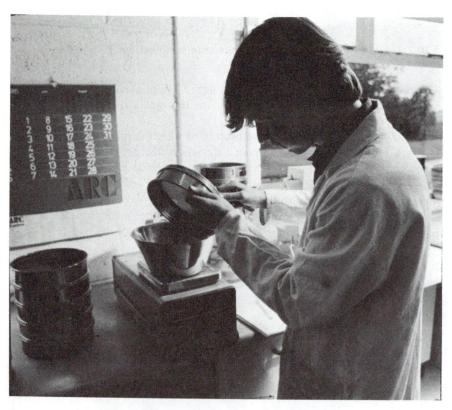

Figure 6.1 Control laboratory. Courtesy ARC Ltd

Quality control: the operation of procedures to maintain product quality at the selected level. This involves decisions and actions by technical, production and general management.

Production control: quality control during production together with operation of preventive measures by production staff.

Quality monitoring: measurement of quality and advising of necessary changes to maintain quality at the selected level—operated by technical and production staff on raw materials and products.

Quality control, production control and quality monitoring are covered in this section.

Quality control of ready-mixed concrete can be divided into three convenient areas [96–98]: forward control, immediate control, and retrospective control. All three include technical and production aspects and use of external data.

6.1 Forward control

Forward control and consequent corrective action are essential aspects of quality control.

Figure 6.2   Site testing. Courtesy Pioneer Concrete (UK) Ltd

Forward control covers:

(i) Materials storage
  Prevention of contamination
  Reliable transfer and feed systems
  Drainage of aggregates
  Prevention of freezing
(ii) Monitoring of qualities of materials
  Visual checks—acceptance/rejection
  Sampling
  Testing
  Certification by suppliers
  Information from suppliers
(iii) Modification of mix designs
(iv) Plant maintenance
(v) Transfer and weighing equipment
  Controlled flow
  Calibrated weighscales
  Separate weighing of cement
  Calibrated admixture dispensers
(vi) Plant mixers and truck mixers
  Blade condition
  Mixer power
  Maintenance.

Examples of properties of aggregates and cements covered by such monitoring are provided in the corresponding materials sections.

6.2 Immediate control

Immediate control is concerned with instant action to control the quality of the concrete being produced or that of deliveries closely following. It covers the following.

(i) Weighing (correct reading of batch data and accurate weighing)
(ii) Visual observation of concrete during production and delivery or during sampling and testing of fresh concrete (assessment of uniformity, cohesion, workability; adjustment to water content)
(iii) Use of equipment at the plant designed to measure moisture content of aggregates, particularly fine aggregate, or the workability of the concrete
(iv) Making corresponding adjustments at the plant automatically or manually to batched quantities to allow for observed, measured or reported changes in materials or concrete qualities.
(v) Inspection at delivery by driver and customer for uniformity and workability; adjustments to workability within company policy.

6.3 Retrospective control

Retrospective control covers

(i) Sampling of concrete; testing; monitoring of results
(ii) Weighbridge checks of laden and unladen vehicle weights
(iii) Stock control of materials
(iv) Diagnosis and correction of faults identified from complaints.

6.4 Quality monitoring

Various methods of analysis are used to compare mean level and uniformity achieved with target levels and to identify need for action. The cusum system of monitoring and analysis is an example of a method applied particularly to strength, but it can also be used to control other properties [99].

Retrospective control can cover any property of materials or concrete, such as aggregate grading, slump, or air content, but is particularly associated with 28-day cube strength because by its very nature it is not a property which can be measured ahead of, or at the time of, manufacture. There is thus always a delay in obtaining results, which implies that action to correct any observed departure from the intended quality will in turn always be delayed.

To partially overcome this, the ready-mixed concrete industry adopts a prediction system using early age tests which rely on the establishment of relationships between the early age strength and 28-day strength [82]. Because these relationships can change, it is necessary also to monitor them continuously.

The early age strength test in most common use is 7-day normal (20 °C) curing. The aspects normally monitored for strength are (i) predicted 28-day strength (mean level, uniformity), and (ii) relation between early age and 28-day strength. The parameter used to measure uniformity is usually the range between consecutive pairs of results, which can be converted easily to standard deviation as follows:

> estimated standard deviation
> = 0.886 × observed mean range of pairs of results

For convenience, in control systems, the formula can be reversed as follows:

> target mean range of pairs = 1.128 × assumed standard deviation

Control systems in common use tend to overestimate standard deviation by about 8% [91] which provides an additional safeguard covering delay in detection of changes in both mean and standard deviation. To allow results from different mixes to be used in the same control system, it is necessary to adjust the results to that of a central 'control mix' on which the system operates. This control mix is usually grade C25 or C30, 50 mm slump, using pc, 20 mm aggregates and without admixtures. All other mixes are correlated

Table 6.1 Relation between design margin and test rate [41]

Monthly test rate for mixes eligible to be included in the control system	Design margin (N/mm^2)
16 or more tests	$2 \times$ s.d.
12 tests	$2.35 \times$ s.d.
8 tests	$2.7 \times$ s.d.
6 tests	$2.9 \times$ s.d.
less than 6 tests	not permitted

with this control mix for design purposes, but only results for mixes using the same cement and aggregates are normally permitted to be used in the control system.

To provide equal assurance under QRSMC regulations [41], ready-mixed concrete producers may opt to test at different rates, but there is a design margin penalty which increases as test rate decreases (Table 6.1).

It should be noted that the test rates in the table cover only tests of those mixes which are eligible to be used in the control systems, and typically a much greater rate of testing actually occurs, covering the full range of concrete produced by each plant. Indeed, in some instances more than one system may be in use at a plant to cover different materials combinations.

As an alternative to increasing the design margin, detection systems are permitted which are faster-acting than the base method described under section 6.5.

6.5 The cusum system of strength monitoring

The following is an adaptation of essential features of the system described in Concrete Society Digest No. 6, 'Monitoring concrete by the cusum system' [100], to which readers are referred for further detail.

A monograph, published by ICI in the mid-1960s, explained the principles of cusum techniques which were applied by Ready Mixed Concrete Ltd to the quality control of concrete. The BRMCA in 1970 and subsequently the Quality Scheme for Ready Mixed Concrete (QSRMC), on its inauguration in 1984, adopted the method as an approved method of quality control in their authorization schemes. BSI published a guide (BS 5703) in 1980, explaining the principles and general application of cusum to quality control.

The cusum system for concrete is used for monitoring trends in mean strength, standard deviation and the relationship between early age and 28-day standard strengths [41, 82, 100].

6.5.1 Principles

Test results are compared with the designed target values, and checks are made to confirm whether they are consistent with the required levels for compliance. In applying the technique, the target value is subtracted from each of the measured results, giving positive and negative differences. These differences when added together for a number of results form a *cumulative sum* (cusum). When this cumulative sum is plotted graphically against the sequence of results, a visual presentation of the trend relative to the target level is produced.

6.5.2 *Control of mean strength*

The technique is easily demonstrated by an example related to concrete strength results. Table 6.2 gives a series of measured 28-day strengths where a producer was aiming for an average of $40 \, \text{N/mm}^2$ to achieve the required characteristic strength of $30 \, \text{N/mm}^2$. Each result is compared with the target value of $40 \, \text{N/mm}^2$, above which the difference is positive, and below, negative. After two results, one $4 \, \text{N/mm}^2$ above and one $2 \, \text{N/mm}^2$ below 40, the cusum is shown as $+4 - 2 = +2$. Between the ninth and eighteenth results it is apparent that a negative trend is occurring. In this case no result has fallen below the characteristic strength of $30 \, \text{N/mm}^2$, but there is a trend towards strength lower than design intentions.

Fig. 6.3 is a graphical plot of this cusum against the result number which visually portrays this trend. The steeper the slope downwards, then the worse is the decline in strength, whereas a consistently upward trend would indicate

Table 6.2 Example of cusum applied to mean strength [100]

Result no.	28-day strength (N/mm²)	Difference from target 40 (N/mm²)	Cusum (N/mm²)
1	44	+4	+4
2	38	−2	+2
3	45	+5	+7
4	41	+1	+8
5	36	−4	+4
6	44	+4	+8
7	35	−5	+3
8	43	+3	+6
9	47	+7	+13
10	34	−6	+7
11	38	−2	+5
12	33	−7	−2
13	42	+2	0
14	36	−4	−4
15	39	−1	−5
16	35	−5	−10
17	41	+1	−9
18	34	−6	−15

MANUAL OF READY-MIXED CONCRETE

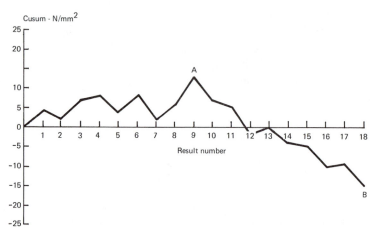

Figure 6.3 Cusum plot of mean strength data from Table 6.2. After Brown [100]

an over-conservative mix design. A horizontal zigzag line would be the ideal situation of the target being achieved. Fig. 6.3 simply suggests that the trend of results is slightly above design level up to point *A*, and then a trend of lower results occurs down to point *B*.

6.5.3 *Monitoring of standard deviation*

When applying cusum techniques to the standard deviation of concrete strengths, use is made of the relationship between standard deviation and the range of successive pairs of results (see section 6.4). The target mean range of successive pairs of results is equal to the standard deviation multiplied by 1.128. Cusum charts are plotted for the differences between the actual range and the target mean range based on the assumed standard deviation from past data. Upward slopes will indicate an increase in mean range and standard

Table 6.3 Target mean ranges [100]

Standard deviation (N/mm²)	Target mean range of strengths (N/mm²)
3.5	4.0
4.0	4.5
4.5	5.0
5.0	5.5
5.5	6.0
6.5	7.5
7.0	8.0

Note: Target mean range = 1.128 × s.d., rounded to nearest 0.5 N/mm².

Table 6.4 Example of cusum applied to range [100]

Result no.	28-day strength (N/mm²)	Range (N/mm²)	Difference from target mean range 5.5 (N/mm²)	Cusum (N/mm²)
1	44			
2	38	6	+ 0.5	+ 0.5
3	45	7	+ 1.5	+ 2.0
4	41	4	− 1.5	+ 0.5
5	36	5	− 0.5	0
6	44	8	+ 2.5	+ 2.5
7	35	9	+ 3.5	+ 6.0
8	43	8	+ 2.5	+ 8.5
9	47	4	− 1.5	+ 7.0
10	34	13	+ 7.5	+ 14.5
11	38	4	− 1.5	+ 13.0
12	33	5	− 0.5	+ 12.5
13	42	9	+ 3.5	+ 16.0
14	36	6	+ 0.5	+ 16.5
15	39	3	− 2.5	+ 14.0
16	35	4	− 1.5	+ 12.5
17	41	6	+ 0.5	+ 13.0
18	34	7	+ 1.5	+ 14.5

deviation, whilst downward trends will reflect a reduction (see Table 6.3 for typical values). An example calculation is shown in Table 6.4, using the data from Table 6.2 and assuming a target mean range of 5.5 N/mm² associated with a standard deviation of 5.0 N/mm².

6.5.4 Significance of trends

Figure 6.3 raises the questions: Is this trend significant? Is it large enough to require action to correct the trend? There is, after all, a high probability that a number of successive results may be above or below target by chance. To overcome this dilemma, a novel method has been devised to confirm when a significant change has occurred. A transparent mask in the shape of a truncated 'V' is placed over the plot, as shown by the dotted lines in Fig. 6.4, with the lead point D over the last cusum result plotted. If the plot remains inside the boundaries of the mask no significant changes have occurred. However, if the plot crosses a boundary, as at C, a significant trend has been detected and action is required. This mask is applied to the plot each time a new result is added and a further check is made.

6.5.5 Design of masks

The geometry of the mask is linked to statistical probabilities, and the need to strike a balance between complex requirements. Significant changes in mean

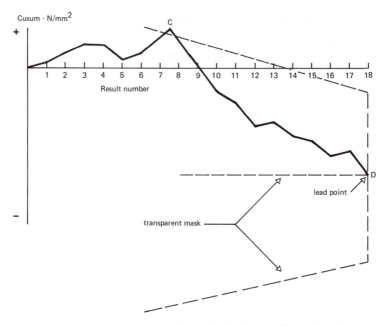

Figure 6.4 Cusum plot of mean strength with mask indicating significant change in mean. After Brown [100]

or standard deviation must be detected as fast as possible, but the system must not be oversensitive so that unreal changes are detected. For example, a mask could be devised which detected a movement away from the target level in only one or two results, but this would detect too many chance situations. Naturally, no matter how many results are involved, there is always some risk of a chance event happening, and this has to be balanced against the risk of missing a real change.

Masks can be designed using either computer simulation techniques or the nomograms available in BS 5703: Part 3. There is no absolute solution, the design being a compromise between the confidence level and speed of detection required, and new proposals are coming forward as experience of the technique develops. The masks from one commercially available system are shown in Fig. 6.5.

Figure 6.5 shows the mask design to be dependent on the target standard deviation, requiring preparation of individual masks for each value of the standard deviation. Alternatively, these individual masks can be superimposed on each other to form a multiple mask, as shown in Figure 6.6.

6.5.6 *Advantages over other systems*

Cusum systems are more sensitive than other systems in detecting changes of the magnitude experienced with concrete production, and reliable decisions

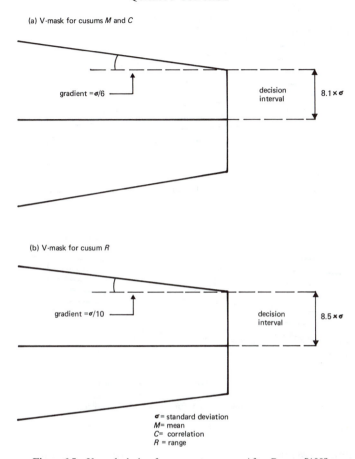

Figure 6.5 V-mask design for concrete cusum. After Brown [100]

can be made on fewer results. Examination of the graphical plot readily identifies the trend of results from the general slope of the graph. Alternations in the slope help to indicate the approximate date when a change in strength or control occurred (e.g. point C in Fig. 6.4). Subsequent investigation can then be concentrated on materials or production procedures occurring at that date. The slope of the graph can be used to determine the magnitude of the change.

Against these advantages there is a slightly increased complexity in processing data compared to other systems. This is of little consequence when the system is computerized.

6.5.7 Cusum for correlation of predicted and actual strength

A cusum system can detect changes in the early to 28-day strength relationship by monitoring the correlation difference (actual − estimated 28-day strength).

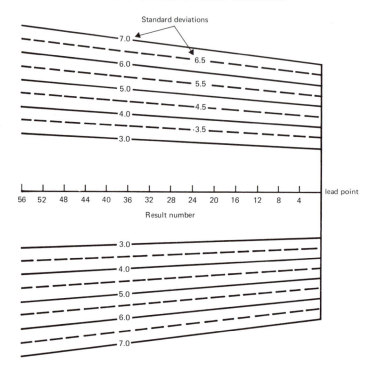

Similar masks can be produced for range and correlation

Figure 6.6   Multiple V-mask for cusum M. After Brown [100]

If the cumulative difference is positive, the prediction system is underestimating and if negative, it is overestimating. Thus an upward slope of the graphical plot represents underestimation and downward, overestimation. Typically, a standard deviation for correlation difference of 2.5 N/mm^2 is adopted.

6.5.8 *Example: cusum in operation* (Table 6.5)

As each early age result is achieved, enter it in the table with its identifying reference data. Consider result 5 in Table 6.5 relating to concrete cast on 10 February. Determine the predicted 28-day result from the appropriate correlation chart and insert it in column 4. Deduct the target mean strength and enter in column 5 (43.5 − 42.0 = + 1.5). Add this difference to the previous cumulative sum in cusum column 6 (4.5 + 1.5 = + 6.0).

Calculate the range between estimated or predicted 28-day strengths between cubes 5 and 4 (43.5 − 38.5 = 5.0) and insert it in column 7. Deduct the target mean range (7.0) in column 8 to give the difference (5.0 − 7.0 = − 2.0). Add this to the previous cusum of range differences in cusum column 9 (2.5 − 2.0 = + 0.5).

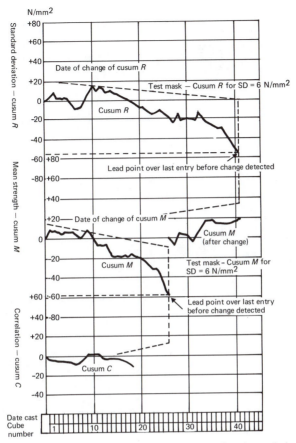

Figure 6.7 Cusum charts for standard deviation, mean strength and correlation for Table 6.6. After Brown [100]

Insert actual 28-day results in column 10 as they occur. Deduct the predicted 28-day strength in column 11 (43.5 − 43.5 = 0). Add this to the previous cusum of correlation difference in cusum column 12 (− 5.0 + 0 = − 5.0).

6.5.9 Plotting cusum charts

Cusum charts are shown in Fig. 6.7 of mean, range and correlation for the data in Table 6.5. Data in columns 6, 9 and 12 are each plotted separately on the vertical axis against the result number on the horizontal axis.

Each chart is examined as every new result is plotted, with the lead point of the appropriate V-mask placed on the graph. No change is apparent for the first 25 results (i.e. the plot does not cross the mask). After result 26 for 10 March is plotted, a mean change is detected. By observing the slope of the plot

Table 6.5 Typical cusum calculations [100]

Initial plant parameters:
Standard deviation 6 N/mm² (Target mean range = 7 N/mm²)
Target mean strength 42 N/mm²

	BASIC DATA			CUSUM MEAN			CUSUM RANGE			CUSUM CORRELATION		
Cube reference	Date cast	Early result (accelerated) (For control mix) N/mm²	Estimated 28-day strength N/mm²	(4) minus target mean strength N/mm²	Cumulative sum of (5) (for mean strength control chart) N/mm²	Range of pairs of values of (4) N/mm²	(7) minus target mean range N/mm²	Cumulative sum of (8) (for standard deviation control chart) N/mm²	Actual 28-day result N/mm²	Correlation difference (10) minus (4) N/mm²	Cumulative sum of (11) (for correlation control chart) N/mm²
(1)	(2)	(3)	(4)	(5)	(6)	(7)	(8)	(9)	(10)	(11)	(12)
1	Feb 5	43.0	54.0	+12.0	+12.0				53.0	−1.0	−1.0
2	8	28.5	38.0	−4.0	+8.0	16.0	+9.0	+9.0	35.0	−3.0	−4.0
3	9	32.0	42.0	0	+8.0	4.0	−3.0	+6.0	41.0	−1.0	−5.0
4	9	29.0	38.5	−3.5	+4.5	3.5	−3.5	+2.5	38.5	0	−5.0
5	10	33.5	43.5	+1.5	+6.0	5.0	−2.0	+0.5	43.5	0	−5.0
6	11	31.0	41.0	−1.0	+5.0	2.5	−4.5	−4.0	39.0	−2.0	−7.0
7	12	33.0	43.0	+1.0	+6.0	2.0	−5.0	−9.0	47.5	+4.5	−2.5
8	15	25.0	34.0	−8.0	−2.0	9.0	+2.0	−7.0	37.5	+3.5	+1.0
9	16	42.0	53.0	+11.0	+9.0	19.0	+12.0	+5.0	53.5	+0.5	+1.5
10	17	24.5	33.5	−8.5	+0.5	19.5	+12.5	+17.5	32.5	−1.0	+0.5
11	17	25.0	34.0	−8.0	−7.5	0.5	−6.5	+11.0	30.0	−4.0	−3.5
12	19	33.5	43.5	+1.5	−6.0	9.5	+2.5	+13.5	43.5	0	−3.5
13	22	32.0	42.0	0	−6.0	1.5	−5.5	+8.0	40.5	−1.5	−5.0
14	23	24.5	33.5	−8.5	−14.5	8.5	+1.5	+9.5	34.0	+0.5	−4.5
15	24	28.5	38.0	−4.0	−18.5	4.5	−2.5	+7.0	35.5	−2.5	−7.0
16	25	33.0	43.0	+1.0	−17.5	5.0	−2.0	+5.0	41.5	−1.5	−8.5
17	26	30.0	39.5	−2.5	−20.0	3.5	−3.5	+1.5	38.5	−1.0	−9.5
18	Mar 1	34.0	44.0	+2.0	−18.0	4.5	−2.5	−1.0	46.5	+2.5	−7.0
19	2	31.0	40.5	−1.5	−19.5	3.5	−3.5	−4.5			
20	2	32.0	42.0	0	−19.5	1.5	−5.5	−10.0			

21	27.5	37.0	−5.0	−24.5	5.0	−2.0	−12.0
22	24.5	33.5	−8.5	−33.0	3.5	−3.5	−15.5
23	32.0	42.0	0	−33.0	8.5	+1.5	−14.0
24	24.0	32.5	−9.5	−42.5	9.5	+2.5	−11.5
25	26.5	35.5	−6.5	−49.0	3.0	−4.0	−15.5
26	24.0	32.5 (35.0)	9.5	58.5	3.0	−4.0	−19.5

Cement content increased by 15 kg/mm³ to return to target mean strength

27	24.5	33.5	−8.5	−8.5	1.5	−5.5	−25.0
28	38.5	49.0	+7.0	−1.5	15.5	+8.5	−16.5
29	37.5	48.0	+6.0	+4.5	1.0	−6.0	−22.5
30	30.0	39.5	−2.5	+2.0	8.5	+1.5	−21.0
31	26.0	35.0	−7.0	−5.0	4.5	−2.5	−23.5
32	39.5	50.0	+8.0	+3.0	15.0	+8.0	−15.5
33	39.0	49.5	+7.5	+10.5	0.5	−6.5	−22.0
34	38.0	48.5	+6.5	+17.0	1.0	−6.0	−28.0
35	32.0	42.0	0	+17.0	6.5	−0.5	−28.5
36	27.5	36.5	−5.5	+11.5	5.5	−1.5	−30.0
37	32.0	42.0	0	+11.5	5.5	−1.5	−31.5
38	32.0	42.0	0	+11.5	0	−7.0	−38.5
39	33.0	43.0	+1.0	+12.5	1.0	−6.0	−44.5
40	33.5	43.5	+1.5	+14.0	0.5	−6.5	−51.0
41	37.5	48.0 (45.0)	+6.0	+20.0	4.5	−2.5	−53.5

Cement content reduced by 20 kg/m³. New target mean 39.0 and standard deviation 4.5.

42	31.5	41.0	+2.0	+22.0	4.0	−1.0	−1.0

it is apparent that the change occurred on or about result 10 on 17 February.

In similar fashion, by using the appropriate range mask, a change in standard deviation is observed at result 41 for 26 February, also originating on or about result 10 on 17 February.

6.5.10 *Action following changes*

When changes in mean strength or standard deviation are detected, the mix should be modified to obtain the appropriate new target levels from the date of detection. It is usual to express such changes in terms of a cement content increment per cubic metre of concrete.

The change in cement content dc is given by

dc	$= 0.75 \times r \times [(DI/n) + G)]$	
where DI	= decision interval of mask	
G	= gradient of mask	(see Fig. 6.5)
σ	= standard deviation	(see Fig. 6.5)
n	= number of results	(see Fig. 6.5)
r	= cement equivalent of 1 N/mm^2 strength (typically 5 kg/m^3)	

0.75 is a reduction factor, sometimes called the anti-hunting factor, included to prevent over-reaction to indicated changes.

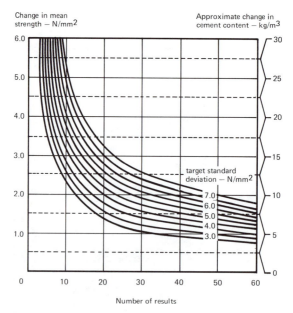

Figure 6.8   Change in mean strength and corresponding change in cement content. After Brown [100]

The number of results (n) between the commencement of the change and the date of detection is sometimes referred to as the run length. Assuming the typical value of $5\,kg/m^3$ of cement equivalent to $1\,N/mm^2$ or substituting any other appropriate value, the change in cement content dc can be calculated.

For ease of use, a chart can be drawn (Fig. 6.8). A similar calculation can be made for changes in the range cusum, bearing in mind that the effect of a change in standard deviation on cement content should take account of the k factor for design margin between target mean strength (TMS) and characteristic strength (C):

$$TMS = C + (k \times \text{standard deviation})$$

Again for ease a chart can be produced, and an example is shown in Fig. 6.9. In this example, cement content changes are based on a k factor of 2. Similarly, change in correlation can be assessed.

Following the detection of change in mean or standard deviation, the cusum calculations and plots for the particular parameter only are recommenced from zero (see plots in Fig. 6.7 related to Table 6.5). When a change in standard deviation is signalled, the new value is used for future assessments of mean and range cusum plots.

When there is a change signalled between predicted and actual 28-day results, then the results subsequent to the indicated point of change are re-

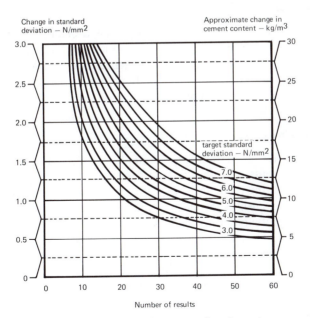

Figure 6.9 Change in standard deviation and corresponding change in cement content. After Brown [100]

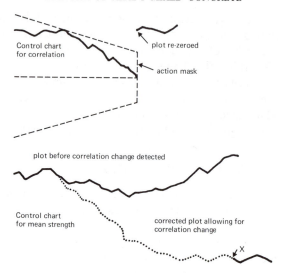

Figure 6.10   Interaction due to change in correlation. After Brown [100]

plotted on the mean cusum, based on 28-day predictions from the new correlation. The effect is illustrated diagrammatically in Fig. 6.10.

6.5.11 *Investigation of the cause of a change*

The cusum allows the approximate date when a change occurred to be identified by observing the change in slope of the plots. This allows attention to be concentrated on differences occurring in materials or production around this time. To a degree, this also helps to avoid false conclusions from events known to have occurred at other times.

6.5.12 *Computerization*

Wherever extensive calculation or tabulation is involved, the use of computers has obvious benefits and this is true for cusum. The only information necessary for routine computer input is cube identification and measured test result. All the cusums for mean, standard deviation and correlation are calculated and plotted automatically. Changes are notified automatically to the operator, and human errors in calculation are avoided.

Significant advantages include the automatic regeneration of tables and plots if the correlation changes between early and 28-day results. Other statistical parameters, such as running mean of four results, can also be printed from the initial data input with no further operator involvement.

Both microcomputers and large mainframes have been successfully used for cusum in quality control of concrete in the ready-mixed industry.

6.6 BRMCA concrete control system

The BRMCA concrete control system [102, 103] has been designed as a comparable alternative to the cusum system which has been incorporated in the QSRMC *Manual of Quality Systems for Concrete* [41]. Its prime advantage is simplicity of use but it is slightly less sensitive than the cusum method. The main differences from cusum are:

(i)   Control of mean strength of the reference mix by use of a simple counting rule
(ii) Direct monitoring of standard deviation every 20 results by simple calculation
(iii) Check of the relation between early and 28-day strength using the counting rule.

The 'counting rule' [103] simply assesses the *number* of results above and below the target value and its use is described in section 6.6.4 on mean strength control.

   The example worksheet (Table 6.6) lists all the control data and summarizes the results of the calculations and action taken. There are five divisions of the worksheet:

(i)   Test data
(ii) Prediction of 28-day strength
(iii) Mean strength control
(iv) Standard deviation
(v)   Early to 28-day strength relationship.

6.6.1 *Initial situation*

Conditions applying in the example at the commencement of the period are indicated in parentheses in the worksheet. They include:

Target mean	$36 \, \text{N/mm}^2$
Cement content of reference mix	$280 \, \text{kg/m}^3$
Standard deviation	$5.5 \, \text{N/mm}^2$

6.6.2 *Test data*

This section (columns *a–e*) summarizes the data, including early age strengths, for the batches of concrete tested over a continuous period of, in this instance, $2\frac{1}{2}$ months.

6.6.3 *Prediction of 28-day strength*

In this section (columns *f–h*), the early strengths are adjusted to predict the 28-day strength in column (*h*) of the reference mix, which in this case (see foot of

Table 6.6. BRMCA Concrete Control System [102]. © BRMCA

	Test data				28-Day strength prediction				Mean strength control			
Cube serial no. (a)	Date cast (b)	Slump (from delivery ticket) (c)	Cement content of mix tested (d)	7-day strength for mix tested (e)	Estd. 28-day value for mix tested (f)	Adjstmt. of mix tested to ref. mix (g)	Estd. 28-day value for ref. mix (h)	Target (T) for ref. mix (j)	Above (T)+; equal, 0; below (T)− (k)	Mean of 10 (l)	Change signalled in mean strength and correction to mean (m)	Change in target (n)
94	14	75	280	21	31	0	31	(36)	−			
5	14	50	260	18	27	0	27		−			
6	16	75	310	30	42	−5	37		+			
7	17	50	230	18	27	+5	32		−			
8	20	50	230	23.5	34	+5	39		+			
9	21	35	250	26.5	38	0	38		+			
100	21	75	280	19.5	29	0	29		−			
1	23	75	280	19	28.5	0	28.5		−			
2	24	50	260	23	33.5	0	33.5		−			
3	27	75	310	30.5	42.5	−5	37.5		+			
4	27	75	310	23	33.5	−5	28.5		−			
5	28	50	260	22.5	33	0	33		−			
6	28	50	230	19.5	29	+5	34		−			
7	Mar. 2	75	280	27	38.5	0	38.5		+			
8	5	75	280	20	30	0	30		−			
9	5	75	280	18	27	0	27		−			
110	7	50	230	19	28.5	+5	33.5		−			
1	8	75	280	21.5	31.5	0	31.5		−			
2	9	75	280	21.5	31.5	0	31.5		−			
3	9	50	260	17.5	26	0	26(29)		−	31.5	−4.5	
4	13	75	310	28.5	40.5	−1.5	39		+			+3
5	14	50	260	20	30	+3.5	33.5		−			
6	15	50	260	20	30	+3.5	33.5		−			
7	15	50	230	17	25.5	+8	33.5		−			
8	19	35	240	22.5	33	+5	38		+			
9	20	75	300	31.5	43.5	0	43.5		+			
120	21	75	300	22.5	33	0	33		−			
1	21	50	250	19	28.5	+5	33.5		−			
2	23	25	260	24	35	0	35		−			
3	26	75	330	26	37.5	−5	32.5		−			
4	27	50	280	26	37.5	0	37.5		+			
5	27	50	250	17.5	26	+5	31		−			
6	29	75	300	26.5	38	0	38		+			
7	30	75	300	27	38.5	0	38.5		+			
8	Apr. 2	50	280	28.5	40.5	0	40.5		+			
9	2	50	250	19.5	29	+5	34		−			
130	4	60	290	19.5	29	0	29		−			
1	5	75	330	28.5	40.5	−5	35.5		−			
2	6	100	315	31	43	0	43		+			
3	6	100	315	23.5	34	0	34		−			
4	9	100	315	27.5	39	0	39		+			
5	10	75	300	26.5	38	0	38		+			
6	10	75	330	26	37.5	−5	32.5		−			
7	12	50	280	23.5	34	0	34		−			
8	13	50	280	27.5	39	0	39		+			
9	16	50	250	17.5	26	+5	31		−			
140	16	100	315	23	33.5	0	33.5		−			
1	17	50	280	27	38.5	0	38.5(35.5)		+			
2	18	75	300	23.5	34	−3	31	33	−			−3
3	19	50	280	19.5	29	−3	26		−			
4	25	50	280	24.5	35.5	0	35.5		+			

			Standard deviation				Early–28-day strength relationship		
Change in cement content and date of change (p)	Cement content of ref. mix (q)	Range of pairs (r)	Sum of 20 ranges (s)	Sum of 40 ranges (t)	Change signalled in standard deviation (u)	Assumed standard deviation (v)	Actual 28-day result for mix tested (w)	Prediction: above actual +; equal, 0; below actual, − (x)	Correlation table reference (y)
	(280)	(3)				(5.5)	29	+	(3H)
		4					23.5	+	
		10					42.5	−	
		5					25.5	+	
		7					36	−	
		1					38.5	−	
		9					26	+	
		0.5	(125.5)	(251)			27.5	+	
		5					35	−	
		4					43	−	
		9					33	+	
		4.5					31	+	
		1					28	+	
		4.5					38	+	
		8.5					34	−	
		3					28.5	−	
		6.5					29	−	
		2					31.5	0	
		0					36	−	
16.3.84		5.5					24	+	
+20	300	10					39.5	+	
		5.5					31.5	−	
		0					33.5		
		0					23.5	+	
		4.5					31	+	
		5.5					44	−	
		10.5					36	−	
		0.5	90.5	216			26	+	
		1.5					36.5	−	
		2.5					38	−	
		5					40.5	−	
		6.5					24.5	+	
		7							
		0.5							
		2							
		6.5							
		5							
		6.5							
		7.5							
		9							
		5							
		1							
		5.5							
		1.5							
		5							
		8							
		2.5							
24.4.84		5	93	183.5	−1.5				
−20	280	4.5			−1.5	4			
		5							
		9.5							

Reference mix C25 20 mm aggregate
pc 75 mm slump

worksheet) is C25, PC, 20 mm aggregate and 75 slump. The adjustments for slump, cement content and early strength are made from relationships established as valid for the particular combination of materials in use at the ready-mixed concrete plant.

6.6.4 *Mean strength control (by counting rule)*

In this section (columns j–q), mean strength is monitored using the counting rule in column (k) and adjustments made to cement content in columns (p) and (q). The individual values of strength in column (h) are compared with target mean in column (j). Results above the target are marked (+) in column (k), results below are marked (−), and results on target, (0).

On each occasion that a sign is entered in column (k) it is examined with the previous 9 signs to check whether 9 (+) or 9 (−) signs have occurred in the group of 10, when an increase or decrease in mean is taken to have been signalled.

For example, for result 113 in column (a), the (−) sign in column (k) in conjunction with the previous 9 signs, all shown boxed, has signalled a significant reduction in mean. This is estimated by calculating the mean (31.5) of the last 10 values in column (h) and recording it in column (l). The reduction in mean is estimated by subtracting the target (36) in column (j) from (31.5) in column (l) and recording (− 4.5) in column (m).

6.6.4.1 *Adjustment to mean strength following signal of a change in mean.* To take account of the normal uncertainty associated with estimation of change of mean from a small number of results (10 in this case), a modification is made to the signalled change of mean.

The adjustment to be made in mean strength is taken as the signalled change, less

$$\frac{\text{s.d. from column } (v)}{\sqrt{10}}$$

In the example worksheet (Table 6.6), at result 113, the observed difference in mean from the target mean is − 4.5 N/mm².

The adjustment to be made is + 4.5 less

$$\frac{5.5 \text{ from column } (v)}{\sqrt{10}}$$

$$= +4.5 - 1.7$$

$$= +2.8(\text{or } +3 \text{ to nearest } 0.5 \text{ above}) \text{ N/mm}^2.$$

which is recorded under the line in column (m) and is achieved by a corresponding increase in cement, as recorded in columns (p) and (q) under the line.

6.6.5 *Standard deviation*

For every value of estimated 28-day strength in column (*h*), the difference from the previous result is recorded in column (*r*).

Every 20 results, these ranges are summed together with the sum of the previous 20. The resulting sum of 40 ranges in column (*t*) is used to check whether any significant change has occurred in standard deviation.

6.6.5.1 *Estimation of standard deviation and corresponding adjustments when a change is signalled.* Standard deviation is calculated from the sum of 40 ranges of pairs of results in column (*t*) as

$$\text{s.d.} = 0.886 \times \frac{\text{sum of 40 ranges}}{40}$$

A new estimate of s.d. is adopted if it differs from the original value by 20% or more.

In the example, at result 141, the sum of the last 20 ranges plus the previous 20 is $90.5 + 93 = 183.5$.

The estimate of s.d. is

$$0.886 \times \frac{183.5}{40} = 4.06 \, \text{N/mm}^2$$

compared with the original value of 5.5 in column (*v*).

Because the difference of $5.5 - 4.06$, which is $1.44 \, \text{N/mm}^2$, say $1.5 \, \text{N/mm}^2$, is greater than 0.2×5.5, i.e. $1.1 \, \text{N/mm}^2$, the new estimate of s.d. is judged to be significantly different from the original and the new value of $4 \, \text{N/mm}^2$, to the nearest $0.25 \, \text{N/mm}^2$, is adopted in column (*v*).

The target mean is changed in column (*j*) by twice the change in standard deviation, i.e. $2 \times (-1.5) = -3 \, \text{N/mm}^2$, and the cement content is changed correspondingly in column (*q*).

6.6.6 *Early–28-day strength relationships*

In this final section (columns *w–y*) of the worksheet (Table 6.6), predictions (*f*) are compared with the actual 28-day strengths (*w*) and use is made of the counting rule method to judge whether the relationship between early and 28-day strength needs adjustment. In the example no violation of the rule has occurred and so the original relation, coded 3H in this example, in column *y*, is assumed to be valid, at least to the date of the latest available 28-day test.

7 Sampling and testing ready-mixed concrete

7.1 Sampling ready-mixed concrete

Compliance tests for fresh and hardened properties of ready-mixed concrete are required to be based on samples fully representative of a batch of concrete, which means obtaining increments throughout discharge. In the case of workability tests, it is thus necessary for the truck to be discharged before a test result becomes available, to confirm what may have been very evident to the practised eye looking into the back of a truckmixer or observing early parts of the discharge. Thus, obtaining and testing a representative sample can hardly be described as useful for acceptance purposes.

To help overcome this problem, the ready-mixed concrete industry proposed and developed a simpler sampling method for slump testing for acceptance purposes only. This method permits a sample to be obtained representing the early part of the delivery after a small proportion has been discharged. The two methods, for compliance and acceptance sampling, are illustrated in Fig. 7.1. Note that for either test method, the test result is the mean of two tests. Because the acceptance sample is not fully representative an increased tolerance on slump is permitted. This method of sampling is not permitted to be used for compliance testing for cubes, for which full representative sampling to BS 1881 is required. For permitted tolerances of BS 5328, see Tables 2.1 and 2.2.

7.2 Effects of non-standard testing on strength

Key aspects of sampling, cube making, curing and testing, selected from the very detailed requirements in BS 1881, are:

  (i) Representative sampling from the truckmixer
 (ii) Remixing of the sample
(iii) Standard cube moulds correctly assembled and oiled
 (iv) Full compaction of each layer
  (v) Maintaining cube in its mould warm and moist overnight
 (vi) Maintaining demoulded cube in water at right temperature until taken to the laboratory

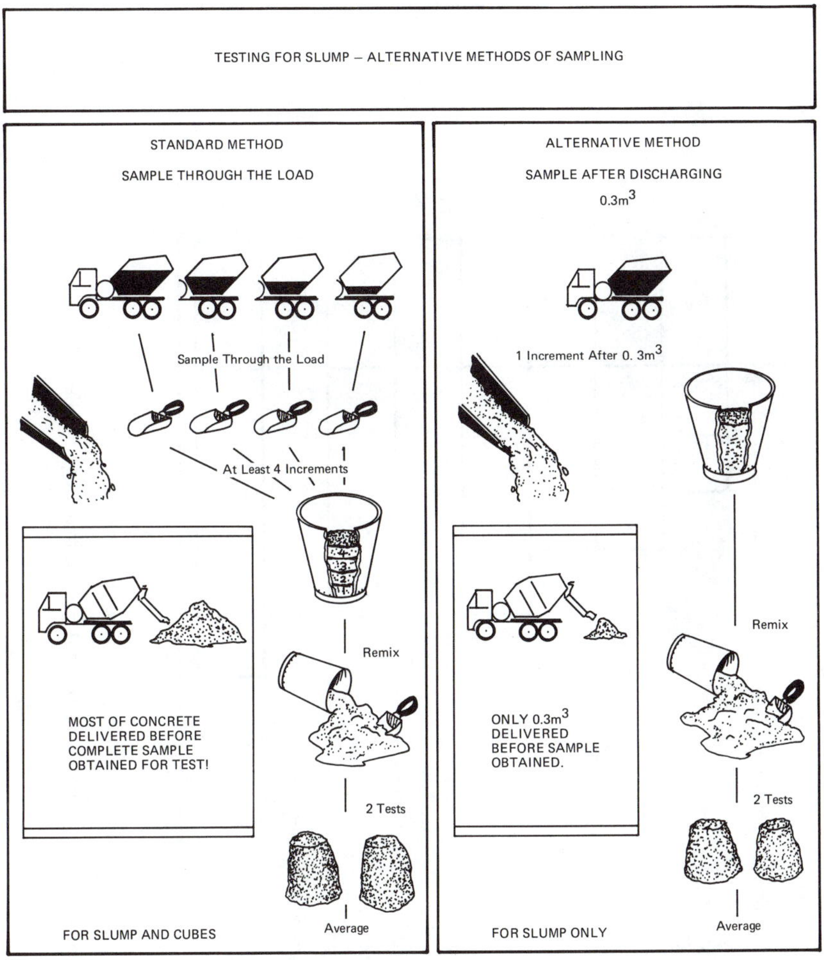

Figure 7.1 Standard and alternative methods of sampling and testing for slump to BS 1881. After Dewar [98]

(vii) Maintaining cube moist, warm and undamaged during journey to laboratory

(viii) As (vi) in laboratory

(ix) Observation of any cube faults, accurate measurement of dimensions, weight and maximum load, and calculation of density and strength

(x) Locating cube centrally and on its side in the machine

(xi) Obtaining correct mode of cube failure in a properly maintained machine.

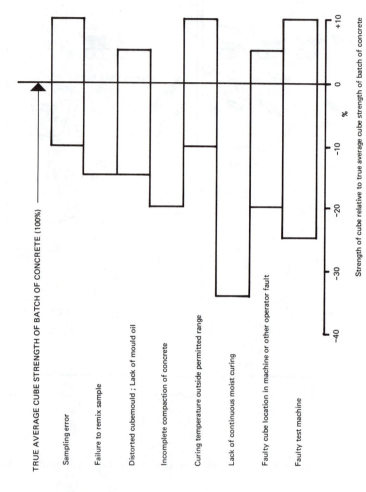

Figure 7.2   Range of effects of deviations from BS 1881 on measured cube strength. Redrawn and adapted from Warren [133]

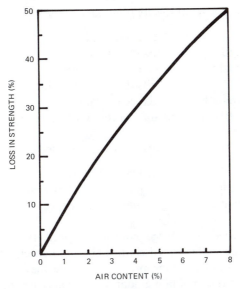

Figure 7.3 Effect of inadequate compaction–relation between entrapped air content and reduction in strength. Based on worldwide data from Popovics [134]

Table 7.1 Effect of an initial period of air-curing on 7- and 28-day strength of concrete, Summarized from J.D. Pateman [104]

Number of days in air*	Loss in strength (%)**	
	7 days	28 days
3	10	4
6	21	8

*Subsequent curing in water at 20 °C until time of testing.
**Compared with control cubes cured for 6 and 27 days in water at 20 °C
Range of control mix strengths:
25–55 N/mm² at 7 days
35–65 N/mm² at 28 days

Effects of some variations from BS 1881 requirements are illustrated in Figs. 7.2 and 7.3 and Table 7.1.

7.3 Simple checks on validity of results

There are a number of simple checks on validity of cube strength results which can be made by producers on their own control cubes and by specifiers or purchasers on cubes made for compliance purposes, as follows.

Measured means of pairs of results from same batch

Cube density	Expected value	$\pm 25 \text{ kg/m}^3$
7-day cube strength	Expected value	$\pm 7 \text{ N/mm}^2$
28-day cube strength*	(i) Expected value	$\pm 10 \text{ N/mm}^2$
	(ii) Above grade value -3	
7/28 day cube strength	Expected value	± 0.15

*The range of $\pm 10 \text{ N/mm}^2$ assumes a standard deviation of about 5 N/mm^2

Measured range of pairs of results from same batch

Cube density	max. 50 kg/m^3
7-day cube strength	max. $15\% \times$ expected mean N/mm^2
28-day cube strength	max. $15\% \times$ expected mean N/mm^2

For medium-sized and large contracts, specifiers or purchasers can maintain simple or sophisticated control charts to check on trends with the aim of detecting and correcting problems, before they escalate into major disasters. Purchasers can obtain guidance in advance on the values to expect from the data bank of the ready-mixed concrete producer.

7.4 Communication

It is vital to develop early communication between specifier, purchaser, producer and test house to compare data and thus continually reduce the risk that differences may remain unrecognized until disaster strikes.

7.5 Simple visual checks on the crushed cube

The mode of cube failure provides the most important visual evidence of the validity of a cube test result in respect of the centrality of cube positioning by the operator and of load application by the testing machine. Naturally, mode of cube failure cannot indicate whether the accuracy of load measurement and its reading and recording by the operator are correct.

A valid cube failure is one for which the cracking is similar on all sides which were vertical in the machine during the test. There should be negligible cracking of the faces contacting the platens of the machine during the test, and any scribe mark impressions from the platens should be central, indicating that the cube was placed centrally in the machine. The trowelled face of the cube should, of course, be one of the vertical sides, because the cube is placed on one side in the test.

Note that with explosive failures, which can occur with high-strength cubes on some machines, the vertical sides may disintegrate completely and parts of the faces contacting the platens may also be lost. This is quite normal, and no significant difference in strength should have occurred.

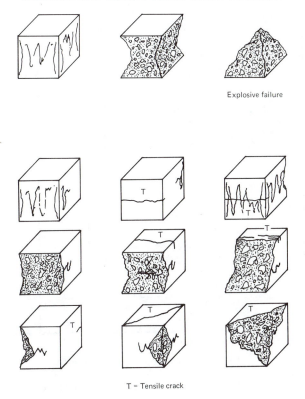

Explosive failure

T = Tensile crack

Figure 7.4   (Top) Normal failures of cubes. (Bottom) Abnormal failures of cubes. After Concrete Society [107]

Sketches of normal and abnormal modes of cube failure are shown in Fig. 7.4. The mode of failure should always be observed and recorded.

One-sided modes of failure are associated with reduced strengths and are caused by one or more of these faults:

(i) Placing cubes off-centre
(ii) Faults in cube moulds
(iii) Locking of the upper platen in its spherical seating before uniform mating has occurred with the cube, as load is being applied
(iv) Rotation of the upper platen in its seating under load
(v) Loose column head nuts or misalignment of machine parts
(vi) Serious lack of planeness of platens or other loaded surfaces.

The importance of the testing machine can be judged from Fig. 7.5 for the results of comparative cube test surveys of grade 2 machines as used by the ready-mixed concrete industry and by test houses in the 1970s. Fortunately, the position has much improved, but is not yet completely fault-free. The

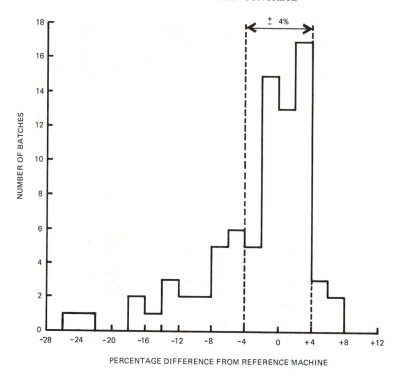

Figure 7.5   Comparative cube tests pre-1970 for Grade 2 machines. After Concrete Society [107]

importance of obtaining a normal mode of cube failure and of recording abnormal failures can be judged from the results in Fig. 7.6 obtained by deliberately crushing cubes off-centre by different amounts.

Considerable research and development by the machine makers in the CTMA, by BRMCA members, by the C & CA, the Concrete Society and BSI have considerably improved the situation, but problems still arise. The importance is stressed of ensuring that machines are regularly serviced and checked at least annually for compliance for load accuracy against BS 1610, and also either with the strain device of BS 1881 or by BCS comparative cube test [105] to check centrality of load application. Note that the BCS comparative cube test also provides checks on other aspects of machine performance and on the operator.

Some users or makers of machines have developed additional techniques to help diagnose faults in machines. For example, Fig. 7.7 illustrates the use of carbon paper impressions obtained under load to provide a simple visual assessment of the area of contact between platens and cubes. It can also be used to detect effects of platens flexing under load or to detect eccentricity of loading.

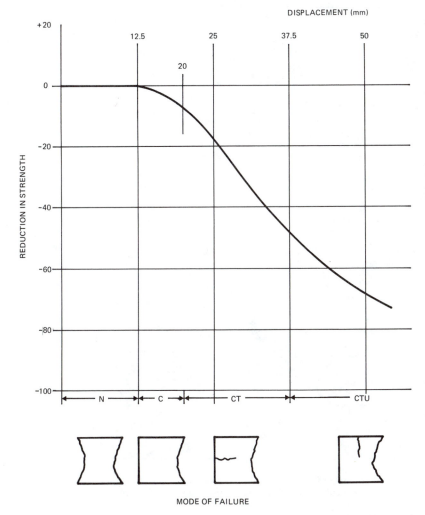

Figure 7.6 Association between mode of cube failure and loss in strength for 100 mm cubes placed eccentrically in a high-quality testing machine *N*, normal mode; *C*, excessive crushing of one face; *CT*, *C* plus horizontal tensile crack in opposite face; *CTU*, *C* plus tensile crack in face against the upper platen. Based on data from RMC Technical Services Ltd

7.6 NAMAS register of test houses

For compliance testing, as recommended by BS 5328 and BS 8000, it is vital that cube tests are made by laboratories accredited by NAMAS (National Measurement Accreditation Service).

A BRMCA Register was established in 1975 to answer the need for high standard laboratories which could be accepted for compliance testing of its

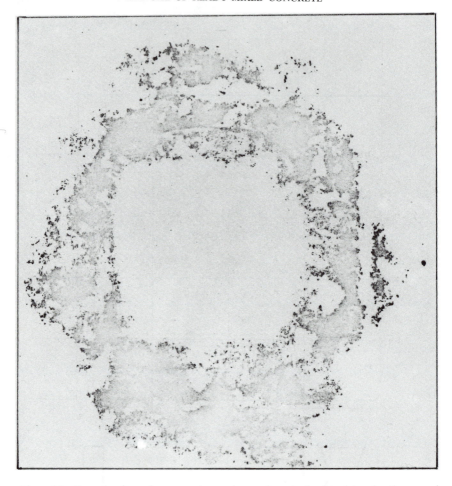

Figure 7.7 Example of a carbon paper impression made under load to determine the area of contact between a cube and an inadequately maintained machine platen

products by the ready-mixed concrete industry. Before this, the precision of testing cubes was often erratic and sometimes very poor (Table 7.2), leading to unnecessary disputes [106, 107].

In 1981 when NATLAS [108] (now NAMAS) was formed with government backing, BRMCA encouraged most of its registered laboratories to join [109]. Laboratories are assessed by chartered engineers against the requirements of BS 1881. NAMAS of course covers a much wider field than cube testing, and it is recommended that NAMAS laboratories are used for any other testing.

Table 7.2 Summary of results of comparative cube tests for 105 commercial testing authorities in 1972. Adapted from Stilwell [110], courtesy of *New Civil Engineer*.

Type of commercial laboratory	No. of laboratories	Percentage of laboratories	
		Deviation from reference laboratory	
		4%–10%	over 10%
General commercial	31	42	19
Contractors	7	29	0
Ready-mixed concrete suppliers	5	20	0
Cement companies	12	42	8
Universities	10	50	10
Polytechnics	6	17	0
Colleges of Technology	30	46	17
Local government	4	75	0
Total	105	42	12

*Differences of less than 4% were judged not to be statistically significant.

7.7 Interpreting test results for strength

7.7.1 *Apparent compliance failures*

The first rule for investigating apparent compliance failures is to establish the validity of the results and the validity of the compliance failure.

Apparent compliance failures fall into three categories:

(i) Valid failures of compliance
(ii) Invalid results associated with testing faults
(iii) Valid results below the specified strength which do not fail any compliance clause.

Results in category (iii) usually only require discussion with the less experienced purchaser of the basis of characteristic strength and an explanation of the compliance rules of BS 5328. The more difficult problem is to differentiate between (i) and (ii) so that appropriate solutions can be found.

7.7.2 *Checking on validity*

The producer has an obvious duty to assure himself that the results are perfectly valid. Unfortunately, in nine cases out of ten investigation shows the results to be based on invalid tests. Even more frustrating, often the problems could have been identified much earlier or possibly prevented altogether (see simple checks on validity in section 7.3).

Obviously, observation of procedures and equipment may help identify whether any deviations from standard requirements are now taking place, but will not necessarily relate to past practice. Sampling and cube-making certificates, as required by BS 1881, should be examined. In more complex cases it may be necessary to painstakingly collect data for complex analysis. As an example, curing may be validated by investigating the ratio of 7- to 28-day strength from site data in comparison with the ready-mixed concrete producer's control data and cement company data.

If early storage on site has resulted in a loss of moisture or exposure to low temperatures, this may have depressed the 7-day strength more than the 28-day strength such that the ratio of 7-day to 28-day strength is depressed significantly, as shown diagrammatically in Fig. 7.8. In a particular case investigated (see Fig. 7.9) it was possible to confirm that there was not only a depression of the 7- to 28-day strength ratio but that this depression increased with duration of site storage before transfer of cubes to the test laboratory, providing indirect but conclusive confirmation of inattention to site curing.

Examination of cube density data is essential, but it is necessary to remember that depression of cube density can be caused by factors other than inadequate compaction or an unrepresentative sample, for example (i) lower cement content than intended; (ii) higher water content than intended; or (iii) higher air content than intended (in the case of air-entrained concrete). Each of these three factors represents a fault in concrete production. Establishing that none of these has occurred may entail chemical analysis (and microscopic analysis for entrained-air content) with all the attendant problems of interpretation of results. It is more valuable to use cube density and

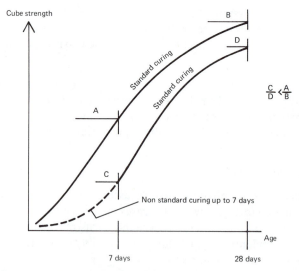

Figure 7.8   Effect of non-standard curing during first few days on strength at 7 and 28 days and on ratio of 7- to 28-day strength (see also Table 7.1)

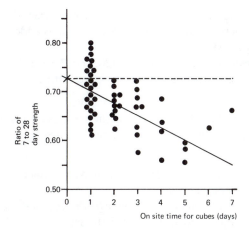

Figure 7.9 Example of effect of non-standard site storage of cubes on ratio of 7- to 28-day strength

visual assessment together as a means of confirming or otherwise the adequacy of compaction.

7.7.3 Action relating to valid compliance failures

When valid compliance failures occur, the specifier, in addition to requiring action to prevent recurrence, has a range of options depending upon the severity of the failure. These are associated with (i) establishing the strength of concrete in the structure particularly at key design locations; (ii) forecasting any significant effect on durability; and, depending on the result of these investigations, (iii) deciding on necessity for remedial action.

It is incumbent upon specifiers in law to seek economic solutions, and indeed most specifiers seek to find the quickest and most economic solution which will enable the construction to proceed swiftly, which is usually in everyone's interest, including that of the specifier's own client.

7.7.4 Establishing the strength of concrete in the structure

The purpose of investigating the structure is to identify, in comparison with structural design requirements, whether or not the in-situ strength is adequate for the actual loading at the age when it is first to be loaded. If there is some reserve strength not utilized in the structural design or if the low strength has occurred at a location of low stress, it may be possible to accept the concrete.

When curing is good and there is adequate time between construction and loading, it may be possible to make allowance for gain in strength between the age of strength assessment and the age of loading. The specifier can call upon the following for assistance:

(i) BS 6089 (Guide to the assessment of concrete strength in the structure)
(ii) Concrete Society Technical Report No. 11, and its addendum on concrete core testing for strength [111]
(iii) BS 1881 (Testing of concrete), Part 120, and Parts 200–208 which advise on the use of core testing, non-destructive testing and interpretation of results.

It is vital to appreciate that concrete in the structure is likely to have a very different strength from the same concrete in standard cubes, even if tested at the same age, because of

(i) Differences in curing, particularly at or near surfaces
(ii) Differences in compaction
(iii) Presence of reinforcement in or near to the concrete sampled
(iv) Sedimentation associated with concrete depth
(v) Temperature stress and internal movements

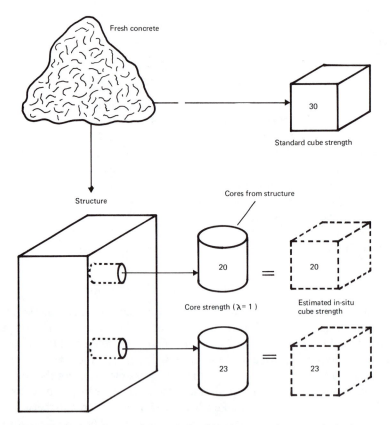

Figure 7.10 Illustration of approximate relationship of compressive strengths in the structure and in standard cubes. Redrawn and adapted from BS 6089

and that, in the case of cores, extraction may create a weakness near the cut surface.

Figure 7.10 shows a typical comparison between cube strength and in-situ strength as provided in BS 6089. The influence of compaction is just as

Table 7.3 Factors to correct in-situ strength for incomplete compaction compared with cubes [111].

Excess voidage* (%)	Strength multiplying factor
0	1.00
0.5	1.04
1.0	1.08
1.5	1.13
2.0	1.18
2.5	1.23
3.0	1.28

*Compared with well-made cubes. See BS 1881: Part 120 and [111] for photographs of cut core surfaces by which assessments of excess voidage may be made.

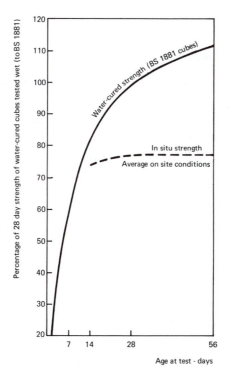

Figure 7.11 Effect of curing upon in-situ strength determined after soaking for 2 days in water. Redrawn and adapted from Concrete Society [111]

important as it was shown to be for cubes. Table 7.3 shows corrections to in-situ strength advocated by the Concrete Society as necessary for different degrees of site compaction. The values are compatible with Fig. 7.3 for cubes. The influence of curing is important too, because it is likely to differ from that adopted for standard cubes, not only at early ages but throughout the life of the structure. It is not only possible but highly likely that substantially lower strength will occur in the structure by 28 days, compared with cubes, and there may often be little gain after 28 days, as shown in Fig. 7.11 for average site conditions. In adverse situations, e.g. unprotected shallow suspended slabs, worse relationships may apply. In favourable situations, e.g. bored piles, strengths comparable with water curing may obtain. Gain in strength with age in structures will depend upon cement characteristics, curing of concrete, carbonation, physical or chemical deterioration and location relative to surfaces.

The Concrete Society has reported [112] work at BRE with present-day cements which indicates that increases of $10-15\,\text{N/mm}^2$, dependent on strength level, can occur from 28 days to 1 year under continuous water curing. In air, the gain is reported to be in the order of $5\,\text{N/mm}^2$ but some of this gain may be associated with surface carbonation and testing cubes in a dry condition. Tested wet to correspond with the testing of cores, the gain may be somewhat smaller.

7.7.5 Interpretation of in-situ cube strength

There are two main ways of interpreting in-situ cube strength data:

(i) Direct comparison with structural design strength
(ii) Conversion to estimates of standard cube strength for comparison with specified cube strength.

The simplest route to follow is (i); (ii) requires agreement of a factor for compaction, and either acceptance of a global correction factor for typical curing or agreement of factors for the effects of age and curing.

For comparison with structural design strength, BS 6089 in effect advises that concrete can be accepted if the in-situ strength is at least $0.80 \times$ specified characteristic strength, or proportionally less if the structural design stress under the actual loading is less than (specified characteristic strength)/1.5, which is the maximum design stress permitted by BS 8110. The Concrete Society Technical Report [111] advises a lower value than BS 6089 as acceptable for estimated in-situ cube strength, as follows:

$$\frac{\text{specified characteristic strength}}{1.5\left(1 - \dfrac{0.12}{\sqrt{n}}\right)}$$

where n is the number of core tests used to determine the mean value.

The difference between the two formulae stems from a difference in purpose. BS 6089 also takes into account the assessment of old structures for new purposes, where little is known concerning the structure and where any long-term gain in strength with age is assumed to be complete. BS 6089 is appropriately more conservative for this application.

Example 1

Specified grade	C30
Age of cores	35 days
Mean in-situ cube strength from 4 cores	25 N/mm^2

(a) Minimum acceptable in-situ cube strength (BS 6089)

$$0.80 \times 30 = 24 \text{ N/mm}^2$$

Conclusion: concrete can be accepted.

(b) Minimum acceptable estimated in-situ cube strength from CSTR/11 [111]

$$\frac{30}{1.5\left(1 - \frac{0.12}{\sqrt{4}}\right)} = 0.71 \times 30$$

$$= 21.5 \text{ N/mm}^2$$

Conclusion: concrete can be accepted.

Estimation of potential (standard cube) strength may be made to CSTR/11 [111] using the following formula, adopting a global factor for curing, an extra factor being needed for any inadequacy of compaction:

estimated potential strength $= 1.3 \times$ estimated in-situ cube strength

The concrete can be accepted if this estimate is not below the specified characteristic strength.

Example 2

Data as Example 1, with additional data:

No correction needed for compaction (based on visual assessment and density tests)
Global correction for curing of 1.30 included in the formula is accepted by all parties.

Estimated potential (standard cube) strength from CSTR/11 [111]:

$$1.30 \times 24 = 31 \text{ N/mm}^2$$

Minimum acceptable estimated potential strength

$$= \text{specified grade}$$
$$= 30 \text{ N/mm}^2$$

Conclusion: concrete can be accepted.

7.7.6 *Use of non-destructive testing*

Non-destructive test methods—BS 1881, [111], [113]—can be used in lieu of core testing or in combination with core testing as described in the previous section to assess in-situ strength. BS 6089 permits the same formula to be used as given for core testing.

7.7.7 *Remedial work*

When assessments of in-situ strength are judged unacceptable, there are a number of options open to the specifier:

(i) Increasing the strength of the concrete
(ii) Strengthening the structure, e.g. by thickening
(iii) Redesigning the remainder of the structure to reduce the load on the weaker elements
(iv) Removal and replacement.

Specifiers are usually loath to resort to the last measure because, unless the defective structure is totally independent of other structures, there may be difficulties in matching appearance or ensuring structural continuity between accepted and replaced sections of concrete.

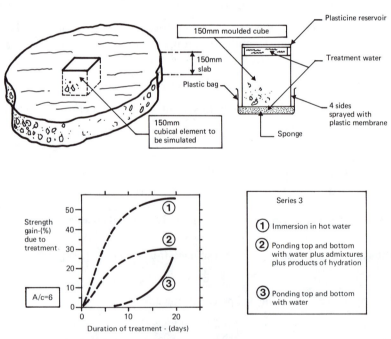

Figure 7.12   (Top) Simulation of treatment of a suspended slab. (Bottom) Gains in strength due to selected accelerating treatments. After Dewar [114]

7.7.8 *Increasing the strength and durability of the concrete in the structure*

It is often possible to rely on natural gain in strength in well-cured or mass construction to improve both strength and durability. In other cases it may be possible to accelerate strength gain, or even to reactivate arrested hydration and to promote a gain in strength which would otherwise not have occurred, or have occurred only over a number of years. A laboratory study simulating practice has confirmed practical observations that significant gains can be obtained [114]. Figure 7.12 illustrates the effect of ponding the top surface together with sealing the under-surface of a 150 mm deep slab, the simulation being made with 150 mm cubes.

It is relatively simple, using non-destructive tests, to confirm at selected points that a gain in strength is being achieved and to stop the treatment when sufficient gain has occurred. Both ponding and steam curing have been used successfully in the UK, and ponding is reported to have been used in Australia [115].

7.8 Checking mix proportions or quantities

Mix proportions may be checked by producer or purchaser as follows:

(i) *Each batch*:
  Observation of batching
  Records of batching (manual or automatic)
  Analysis of fresh concrete
  Chemical analysis of hardened concrete
  Indirect checks on water/cement ratio (slump, strength of cubes)
  Indirect checks on cement content (strength of cubes)

(ii) *Overall*:
  As for each batch above
  Materials stock reconciliation.

7.8.1 *Analysis of fresh concrete*

Several major problems which have led to only limited adoption of methods of fresh-concrete analysis include the necessity for skilled operators, the expense and complexity of tests, sampling inaccuracy, imprecision of the tests, time to obtain a result, and inapplicability to small sites.

A comprehensive BSI draft for development for fresh analysis (DD82) describes several methods in detail and also lays down requirements for assessing precision to ensure maximum accuracy from each method. However, the main difficulty [116] is the introduction of an unknown bias because of assumptions which cannot be confirmed as true for any given sample. For example, physical methods for assessing cement do not discriminate between silt and cement. An assumption has to be made of the relative proportions. In the case of chemical methods based on lime determination, an assumption has to be made concerning the relative proportions of lime in the cement and aggregates.

Fresh analysis has been used on a few large contracts as a compliance test, but only rarely as an acceptance test because of the delay in obtaining results. It does have one important use: assessment of mixing uniformity of plant mixers or truckmixers, where relative comparisons are of interest rather than absolute values.

7.8.2 Chemical analysis of hardened concrete

Chemical analysis is beset by similar problems to fresh analysis, with the additional greater difficulty in ensuring representative samples, greater problems of test imprecision, increased time for results, and higher costs. As a result, chemical analysis of hardened concrete is not normally acceptable as a compliance test, and is often the cause of further argument when used later in the event of dispute. Concrete Society Report 32 indicates poor precision even under ideal conditions [117].

Some indication of the variation in chemical analysis which can occur in practice is given in Table 7.4 [118]. The cubes analysed were straightforward, without the complexities relating to blended cements, ggbs and pfa which increase the potential for unreliability still further. Its prime value is for diagnostic purposes, where the interest lies in the presence or absence of materials, and as an indication of the order of magnitude of proportions, not exact values.

Table 7.4 Variation in reported aggregate/cement ratio from chemical analysis of hardened concrete without complexities, by eleven testing authorities [118]

Testing authority	A/C ratio by mass
Cubes* as prepared	8.0
1	10.7
2	5.9
3	8.1
4	8.6
5	7.2
6	6.9
7	5.4
8	8.6
9	8.9
10	8.0
11	8.8

*Individually hand-batched and mixed.

Where chemical analysis for included chloride content of samples taken from structures is concerned, interpretation is made difficult because of the possibilities of migration of chlorides within the concrete and ingress of externally applied chlorides, such as de-icing salt.

7.9 Checking the quantity of concrete

The quantity of ready-mixed concrete produced or delivered can be checked in a number of ways by the supplier, purchaser, specifier and inspector for trading standards or quality assurance.

(i) *Individual batches*:
   Observe batched weights of all materials
   Obtain gross and tare weights of the delivery vehicle over a stamped weighbridge
   Measure volume of (a) the excavation or forms before concreting; (b) plastic concrete in place; (c) hardened concrete construction
(ii) *Summation of batches*:
   As above for individual batches
   Check on materials opening and closing stocks and materials delivered in comparison with delivery tickets and production records for concrete produced.

Each of the methods involving weight relies on a knowledge of the plastic density of the concrete for conversion from weight to volume:

$$\text{volume of concrete produced} = \frac{\text{weight of all materials batched}}{\text{plastic density of concrete}}$$

As described in BS 5328, the only standard method for judging the volume of a batch of delivered concrete is by weighing the vehicle over a weighbridge before and after delivery, together with a plastic density test made on that delivery, in accordance with BS 1881. Assumption of a value for plastic density or estimations from mix proportions and relative densities of materials can lead to significant errors and should only be used for guidance.

In the special case of semi-dry concrete, a method for determining density is given in BS 1881: Part 129.

7.9.1 *Measuring construction volume*

Sources of underestimation of volume of concrete required or used include the following:

(i) Wastage and spillage
(ii) Water leakage, evaporation or absorption by forms or subgrade
(iii) Bowing of wall forms, flexing of slab soffits

(iv) Filling of gaps
(v) Over-excavation
(vi) Unevenness of subgrade.

7.9.2 *Measuring volume to be concreted*

Direct volume measurement may be satisfactory, provided account is taken of the sources of error given above, and a check is made after concreting to confirm that the final top surface level does not differ from that assumed and the forms have not moved. For example, if the centre of a 150 mm slab deflected downwards by 5 mm during concreting but the upper surface was laid level, an increased volume of 3% would be required at the centre, decreasing towards the edges. In fact, the centre is more commonly cast proud relative to the edges, which would increase the volume required still further.

Reliance should not be placed on construction drawings—all measurements should be made on site. The volume of concrete in bored piles, particularly those where the pile is withdrawn, can often be underestimated because of oversize excavation.

7.9.3 *Measuring volume of plastic concrete after finishing*

This method is useful for paving or slabs, provided depth is measured at sufficient points over the area concreted to enable an accurate average depth to be measured.

7.9.4 *Measuring volume of hardened concrete*

This can be an accurate method provided all significant dimensions can be measured. The most difficult problem is the concrete slab, where the underside is inaccessible. No reliance should ever be placed upon measurements made at the edges alone, because the edge cannot be guaranteed to be representative of the depth elsewhere. Sets of levels of the base before concreting and the top surface after hardening are to be preferred, but again, the sources of errors indicated earlier need to be considered.

In certain cases, holes may be permitted to be drilled to obtain direct measurements of depth. There are methods of non-destructive testing which enable depth of slabs to be measured, but it is important for calibrations to be confirmed by making some test drillings for direct depth measurement.

7.10 Variation in density and yield

It is important to establish relationships between plastic density and mix (i.e. cement content) to ensure correct batch weights for each cubic metre of concrete, for each type of concrete and set of materials in use. Normal variations with properties of materials do not have a significant influence on either density or yield. However, it is important to take account of variations in entrained air content and in density of lightweight aggregates which may significantly affect concrete density and yield.

Part 2

PRACTICE

8 Production, delivery and quality assurance

8.1 Production methods [119]

Ready-mixed concrete is produced in the UK using many different combinations of each of the elements of the process: handling the incoming materials; batching the constituents; mixing the concrete; loading the delivery vehicles. The industry has not settled on a 'best way', and choice has been based on economics related to perceived markets for each particular plant. As the shelf life of the product is so short, the production units (the plants) have had to be located convenient to potential markets.

To design a plant to produce ready-mixed concrete necessitates having realistic assessment of the likely size of the market and the types of concrete used locally, as well as:

   (i) Types of materials
  (ii) Storage capacity
 (iii) Processing sequences
  (iv) Truckmixer capacity and throughput
   (v) Planning requirements
  (vi) Quality control requirements
 (vii) Regulatory requirements
(viii) Duration of operation
  (ix) Health and safety regulations.

8.1.1 *Material types*

The different types of cements for which provision has to be made will depend upon the types of concrete likely to be required in the market area, whereas the types of aggregates will be related to what is readily available: gravel, land-borne or marine; limestone; granite; sand; crushed rock fines. It is normal practice to stock 20 mm and 10 mm maximum-size coarse aggregate, and also 40 mm, depending upon local practice.

A twin-silo plant that stocks pc and ggbs or pfa will be able to produce ordinary concrete and some grades of sulphate-resisting concrete, but not, for example, rapid-hardening concrete, for which a silo would need to be emptied and restocked with rapid-hardening cement.

8.1.2 *Storage capacity*

Handling incoming materials in bulk is fundamental to the economic production of ready-mixed concrete. The current legal weight limit on vehicles, based on the 10-tonnes maximum axle load, is a major factor influencing the delivery costs of cements and aggregates. Currently, cement tankers deliver up to 20 tonnes of cement and, when concrete production demand rises, the ready-mixed concrete producer relies on the cement maker's tanker fleet to keep the silos stocked, although changes in cement pricing have meant that it can be advantageous for the ready-mixed concrete company to collect the cement ex-works. Modern tippers carry up to 24 tonnes of aggregates. The design of the storage elements, cement silos and aggregate storage bins is obviously influenced by the bulk delivery vehicle capacity. Where plants are situated next to sources of aggregate supply, such as quarries, gravel pits, railheads or wharves, then storage facilities for the aggregates at the ready-mixed concrete plants do not need to be so large. Stabilizing the moisture content of the aggregate, particularly the sand, can be an important factor in determining the aggregate storage capacity and layout.

8.1.3 *Processing sequence*

A plant may be designed for (i) batching; (ii) batching and mixing; or (iii) batching, with a mixing option. The maximum batch size is a major factor in determining the rate of production of a plant, but few plants are designed to load a truckmixer with materials in a single batch.

The accurate weighing of solid constituent materials has been a fundamental requirement for producing quality concrete for many years, but the degree of automation built into the batching process has only in recent years shown a marked increase. The reliability of automatic controls has improved and computerization has enabled more sophisticated adjustments to be incorporated. The commonest weighing equipment continues to be the balanced mechanical lever system, although load cell weighing is used. Some continuous weighing systems are still in use. Controlling the discharge from storage and weigh hopper (Fig. 8.1) is in the main through gates operated by compressed air cylinders.

With increasing emphasis on calibration, there is a need to be able to carry out the regulatory routine and specialist checks of the weighing process without too much difficulty, and this is having its influence on plant layouts. The use of flow meters for water measurements almost universal, although in some plants the water is weighed. The use of dispensers to measure admixtures is almost universal. Again the need to be able to calibrate them easily is having an influence on their design and installation.

Because modern truckmixers are able to thoroughly mix a wide range of concrete mixes, the justification for including a mixer in the processing

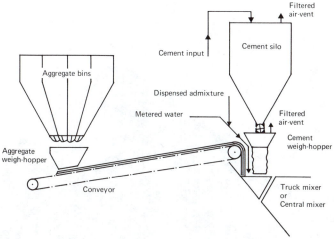

Figure 8.1   Weigh hopper system. Redrawn and adapted from Crowther [144]

sequence is often influenced by the local market demand for lean mix concretes. These semi-dry concretes are frequently collected in tippers, and so a mixer has to be included in the plant design.

In general, plants in which all the concrete has to be processed through the mixer are only situated in urban locations. A compromise is to have two production streams, with a 'dry-leg' going straight into the truckmixer and a 'wet-leg' going through a mixer. Such 'wet-leg' mixers are usually of limited capacity.

As well as adding to the capital costs a mixer will certainly increase the energy requirements of a plant and the maintenance costs, so the decision whether or not to include or omit a mixer in the production process can be a vital one in determining its commercial viability.

In the UK the most common method of producing ready-mixed concrete is to load the batched materials into a truckmixer, mixing them together by rotating the drum. The most important factor in determining within-batch uniformity is the method used to charge the mixer. Practices have evolved to suit the characteristics of the local materials, but if concrete is not well mixed the first thing to examine is the loading sequence and the procedures used to add water. The options available for loading a mixer are:

   (i) Ribbon loading
  (ii) Cement-last loading
 (iii) Sandwich loading
 (iv) Slurry mixing
  (v) Single batching
 (vi) Multiple batching.

As well as seeking uniformity of the mix, the loading sequence must be such as

Figure 8.2 Urban plant with central mixer. Courtesy RMC (UK) Ltd

to avoid a headpack or the formation of cement balls. Headpacks occur when the sand or sand and cement is packed into the head of the drum, remains there and is not mixed into the concrete. This can be avoided by ensuring that coarse aggregate and/or water gets down to the lower end of the drum. Cement balls are accumulations of cement and sand rather than purely cement, and are more likely to occur in batches mixed for only a small number of revolutions.

(i) Ribbon loading: This involves blending the coarse aggregate, the fine aggregate and the cement throughout the majority of the loading cycle. The method of charging the water is varied to suit the materials.

(ii) Cement-last loading: The aggregates are loaded, then the cement is added.

If the cement is added without turning the truckmixer drum and mixing water is not added nor mixing done until arrival at the site, then very little cement becomes wetted. This can be advantageous when the concrete has to be transported long distances.

(iii) Sandwich loading: In this loading the cement is sandwiched between equal increments of ribboned aggregates and water.

(iv) Slurry mixing: Water is loaded first followed by the cement and a slurry formed for a short period before the aggregates are ribbon loaded.

(v) Single batching: All the materials for one load are batched and loaded in one cycle.

(vi) Multiple batching: The materials for a load are batched in two or more cycles dependent upon the capacity of the weighing systems.

8.1.4 *Truckmixer capacity and throughput* (see 10.1)

The number of truckmixers operating from a plant is usually the critical factor in determining the throughput, and has to take into account both truckmixer capacity and radius of operations. The average number of truckmixers operating from plants in the UK is three and the average radius of operations is 8 km. When the demand for concrete is high, the number of truckmixers is increased, and when demand is low, sales are sought further away from the plant.

Figure 8.3 Low-level batching plant with truckmixer. Courtesy Redland Readymix Ltd

8.1.5 *Planning requirements*

The requirements of planning authorities for the approval of sites for ready-mixed concrete plants vary, dependent upon the location. Environmental conditions are different in a quarry from those pertaining in an urban industrial estate, but producers have to meet the requirements of the Environmental Protection Act and the discharge requirements of the National Rivers Authority (NRA).

8.1.6 *Quality control requirements*

The proliferation of methods currently being used to specify concrete means that a responsible ready-mixed concrete producer who wishes to meet all market demands must have a very comprehensive quality system. The requirements for sampling, making cube specimens, curing and testing them under standard conditions involves the setting up of a laboratory where the test data can be collated and new mixes designed. The cost of quality control facilities has encouraged the grouping of plants within a company into areas, with the quality control being managed from one technical centre. The laboratory and mobile sampling and testing facilities and staffing, needed to monitor the constituent materials and to ensure that mixes supplied comply with the orders received, form a key part of any company's quality system, and the cost of setting up a laboratory and control system is a major item.

8.1.7 *Regulatory requirements*

British Standards lay down basic standards covering the production process which have to be met:

    BS 1305: Batch type concrete mixers
    BS 1881: Testing concrete
    BS 5328: Concrete

Since the 1950s trade practice has evolved its own requirements covering production and delivery of ready-mixed concrete, to ensure that:

(i) Each plant has efficient equipment provided and well maintained to ensure the capability to produce and deliver a range of concrete mixes to the standards specified in BS 5328 in respect of quality and quantity
(ii) Methods of material storage and handling, and concrete production and delivery are used which minimize the risks of non-compliance of the product in terms of quantity and quality.

The current requirements for good practice are set out in the Technical Regulations of the Quality Systems for Concrete, published by QSRMC, with which all members of the Scheme have to comply.

8.1.8 *Duration of operations*

The estimated time a plant is to be in operation is a critical factor in determining the type of plant to be installed. Few plants are truly mobile, but transportable plants are quite common, and many ready-mixed concrete producers now set up plants on major construction sites for the duration of the job. Whatever the period of time for which the plant is to be used, considerable expenditure on foundations, services, erection and dismantling, access roads and paved areas, wash-downs and effluent disposal will be required. No matter how relatively short the time a plant is to be in operation, the product has to be of the correct quality, so the standards of good practice remain the same.

8.1.9 *Health and safety*

For ready-mixed concrete plants the Health and Safety Executive have recommended that particular attention needs to be paid to the following:

  (i) *Conveyor guarding*: Adequate guards on conveyors and conveyor drives; positioning of the bottom pulley to facilitate clearing of spillage, safe access for correction of tracking and adjustment of belts when in motion
 (ii) *Access to mixers*: Interlocking of access points on mixers, so as to avoid hand and arm injuries
(iii) *Emergency stops and lockout procedures*: Sensible positioning of belt isolators: system of lockouts for ensuring that plant is isolated from the power supply when guards have to be removed for maintenance
 (iv) *Traffic control*: Routeing of incoming cement tankers, aggregate tippers and outgoing truckmixers to minimize reversing; ensuring that vehicles and pedestrians are separated; use of audible warning systems
  (v) *Inspection points*: Provision of grids and pulpits at ground hopper tipping points
 (vi) *Inspection system*: Providing a safe system for the fresh concrete to be visually checked during mixing
(vii) *Safe access*: Provision of edge protection and safe access to all parts of the plant for maintenance as well as operating staff
(viii) *Silos and tanks*: Provision of safe entry into silos and tanks; provision of relevent safety equipment, including lighting and anchorages
 (ix) *Security*: Security on the site against access by members of the public.

8.2 Organizing production and delivery

Organization within companies, for operating and controlling production and scheduling deliveries to sites, varies from company to company. Some companies' operations are controlled at depot level, whereas a number of the larger and medium-sized companies have centralized shipping offices covering

Figure 8.4   Central shipping office. Courtesy C & G Concrete Ltd

particular markets (Fig. 8.4). Whilst the requirements lend themselves to computerization, the economics are influenced by geography and market conditions. Part of any company's organization is a very strict credit control system, made necessary by the short period between enquiry/order and delivery, and the inability to reclaim the delivered products.

Management systems have been developed using computers to improve productivity levels. The extent of computerization [120] has varied between

BATCH CARD

MIX CODE ....   Target Slump ....

Dry batch weights for 1 cubic metre	Moisture content of coarse aggregate
kg	
Cement .... kg	40 mm aggregate ....%
40 mm aggregate .... kg	
20 mm aggregate .... kg	20 mm aggregate ....%
10 mm aggregate .... kg	
Sand .... kg	10 mm aggregate ....%
Total water ....litres	
....	

| Quantity | Cement | CUMULATIVE BATCH WEIGHTS | | | | | Water | |
		40 mm	20 mm	10 mm	Sand			
m^3	kg	kg	kg	kg	M/C	kg	litres	
½					6%			
					8%			
					10%			
1					6%			
					8%			
					10%			
1½					6%			
					8%			
					10%			
2					6%			
					8%			
					10%			
2½					6%			
					8%			
					10%			
3					6%			
					8%			
					10%			

Figure 8.5 A basic batch card

companies, from virtually complete computerization of every function, to the use of small personal computers for individual element, e.g. batching instructions print-out in place of manually prepared batch-books. An example of a basic batch card is given in Figure 8.5.

Complete software and hardware packages are available [121] covering:

(i) Accepting orders from customers
(ii) Executing deliveries to customers including batch/mix control
(iii) Scheduling constituent materials to plants
(iv) Administration and invoicing.

8.3 Delivery

8.3.1 *Truckmixers* [122]

Ready-mixed concrete is not only a product, it is a service, and each year about 20 million cubic metres of concrete are delivered in truckmixers [118]. The truckmixer has developed since the late 1940s from a mobile site mixer into a specialized vehicle capable of mixing, delivering and distributing concrete in a very economic manner. Indeed, in the viability of ready-mixed concrete depends on the efficient utilization of the specialized truckmixer fleet.

8.3.1.1 *Mixing.* The versatile truckmixer can function in the production process in three basic ways. (i) As a mixer at a ready-mixed concrete plant: the truckmixer is loaded with the dry batched materials together with water and mixing is completed at the plant. During transit, the drum can be revolved slowly to keep the mix agitated. On arrival at site, the drum is rotated at mixing speed for a few minutes to ensure complete remixing, before discharge. (ii) As a mixer at the site: the truckmixer is loaded with the dry batched materials at the plant, which are then transported to the site. On arrival, mixing water is added and the mixing completed before discharge. (iii) As an agitator: the truckmixer is loaded with mixed concrete from a mixer at the plant. During transit the drum can be revolved slowly to keep the mix agitated and at site the drum is revolved at mixing speed to ensure complete remixing before discharge.

8.3.1.2 *Delivery.* Although truckmixer vehicles available in the UK range from 2 to 9 cubic metres capacity, the majority in use have a load capacity of 6 cubic metres of concrete (Fig. 8.6). With a 'shelf life' of only a few hours, ready-mixed concrete is very much a 'local delivery' service, the average distance from the depot to point of delivery being about 8 km, varying between town and country. Long deliveries are possible, but the economics need careful scrutiny.

8.3.1.3 *Distribution.* On arrival at the site, the standard truckmixer can:

Figure 8.6 A 6 cubic metre truckmixer. Courtesy Ritemixer

(i) Mix the concrete and, with measured water addition, adjust the characteristics of the plastic concrete
(ii) Using a fitted chute and extensions, discharge the concrete direct
(iii) Control the rate of discharge to suit the placing requirements on site.

8.3.1.4 *Type of vehicles.* The mixer unit, mounted on a commercial vehicle chassis, consists of an abrasion-resistant steel drum inclined at about 16°. There are blades inside the drum to mix and agitate the ingredients when the drum is rotated in one direction, and to discharge the mixed concrete when the drum is reversed. The power for turning the drum can come from the main truck engine, using a hydraulic power drive to the drum; a separate (donkey)

SHIPPING SHEET

Plant:

Date:

WEATHER

TEMPERATURE °C
Max   Min

Order	Code	Radial	Customer Name Site Location	Quantity Ordered	Quantity Delivered	Mix	Slump	Aggregate Size	Aggregate Type	Cement	Admixture	Remarks
	A											
	B											
	C											
	D											
	E											
	F											
	G											
	H											
	J											
	K											

Daily Delivery Plan

Truck Number	7.00	7.30	8.00	8.30	9.00	9.30	10.00	10.30	11.00	11.30	12.00	12.30	1.00	1.30	2.00	2.30	3.00	3.30	4.00	4.30	5.00	5.30	6.00	6.30	7.00	Total

Total Deliveries

Figure 8.7 A basic shipping sheet

engine, with hydraulic drive; or a separate engine with a mechanical coupling to the drum. The cost of a complete unit (chassis, cab and mixer unit) is about £56 000 (1991 prices). It is possible to change the drum when it gets worn, the rate of wear being very dependent upon the abrasive effects of the aggregates, but most companies operate on a life of over four years for the mixer.

There are three major producers of truck-mounted mixer units in the UK, but manufacturers from other countries are also selling units in the UK.

8.3.1.5 *Methods of operation.* The average number of truckmixers operating out of one depot in the UK is three. However, trucks can easily be moved from one depot to another to meet peak demands. The average number of deliveries made by a truckmixer is five per day. Where a company has a group of depots in an area, rather than allow each individual depot to control its vehicles, 'central shipping' has been found to lead to higher utilization of the trucks. An example of a basic shipping sheet is given in Fig. 8.7. Various sizes of truckmixers in a fleet can assist in economic scheduling of deliveries but in the main it is usual to standardize on one size.

There are about 3000 truckmixers operating in the UK. The majority of these trucks are operated by 'owner-drivers'. The owner-drivers are not company employees, but are contractors with exclusive contracts to carry one company's product on a hire or reward basis. There is a close contractual and financial relationship between the driver and the company. The truck chassis is either purchased or leased by the driver, but the mixer unit normally remains the property of the company.

Some ready-mixed concrete companies own truckmixers and employ 'company' drivers either exclusively or alongside 'owner-drivers'.

8.3.1.6 *Regulations.* Each driver of a standard truckmixer has to hold an HGV Class II Licence. If he is an owner-driver he also has to hold a Certificate of Professional Competence or be exempt from its requirements before he can obtain a Standard Operator's Licence. The taxation and insurance costs of a truckmixer are currently about £3000 a year.

8.3.1.7 *Why a specialized vehicle?* The justifications for a specialized vehicle for delivering concrete are that it can:

(i) Safely convey a whole range of concrete mixes
(ii) Agitate the mix in the time between mixing and discharging
(iii) Allow the concrete workability to be adjusted, by adding water if necessary
(iv) Distribute the concrete by driving on and around the site and by means of an extended chute control the height and radius of discharge.

Non-agitating vehicles can be used to deliver mixed concrete that is not prone to segregation.

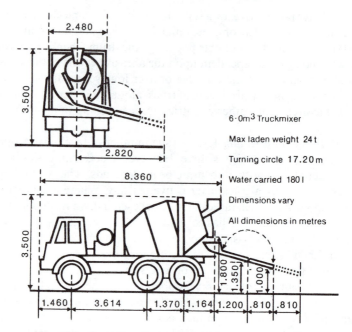

Figure 8.8 Typical truckmixer dimensions. Courtesy Ritemixer

8.3.1.8 *Making the most of truckmixers.* For the purchaser of ready-mixed concrete to get full benefit from a truckmixer delivery of concrete requires:

(i) Good access and room for the vehicle to manoeuvre. A truckmixer is about 3.5 m high, 2.5 m wide (Fig. 8.8), and can weigh as much as 24 tonnes. Truckmixers have some cross-country capability, but they are basically road vehicles.

(ii) The delivery to be planned to suit the handling conditions on site. Good liaison with the producing plant is essential as 6 cubic metres of concrete can be discharged in under 5 minutes.

(iii) Quick turn-around of truckmixers at the site. The supplier normally charges for delays on site.

8.3.2 *Tippers*

A significant amount of concrete is collected direct from ready-mixed concrete plants in tippers.

If the plant has a wet-leg then discharge of the mixed concrete is made direct into the tipper. When there is not a central mixer at the plant the constituents are batched into a truckmixer, mixed and discharged from the truckmixer into the tipper, using a ramp if necessary, to get the discharge chute above the sides of the tipper.

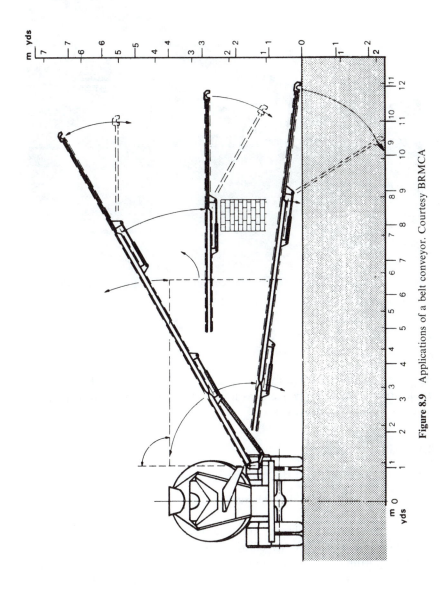

Figure 8.9 Applications of a belt conveyor. Courtesy BRMCA

Figure 8.10   Conveyor mounted on a truckmixer. Courtesy Tarmac Topmix Ltd

8.3.3 *Conveyors*

A few truckmixers have conveyors [123] attached to them (Fig. 8.10). The applications of a belt conveyor are shown in Fig. 8.9. A second truckmixer can be used to feed the conveyor.

8.4 Quality assurance

To have assurance about the qualities of concrete requires that:

(i) The qualities required of the fresh and hardened concrete are objectively defined
(ii) The production delivery and site processes are carried out under controlled conditions which allow the whole process to be able to be traced from the constituent materials right through to the cured concrete
(iii) All testing carried out in the control and compliance testing is valid.

Ready-mixed concrete production and delivery is only part of the QA process, but if the concrete is produced under conditions complying with BS 5750 (Quality systems) and BS 5328 (Concrete), then the degree of assurance is increased.

The ready-mixed concrete producer can only supply what is ordered, because the contract to supply only exists between the purchaser and supplier, so that the ready-mixed concrete producer is dependent upon the interpretation of the specification by the person who eventually orders the concrete.

The concrete producer completes his part of the contract having produced the concrete and delivered it down the discharge chute. He has no control over how the concrete is placed, compacted, finished, cured or protected. Production control testing by the ready-mixed concrete producer is required to be in accordance with the agreed standards. Unless the compliance testing carried out by the purchaser is in accordance with the same agreed standards it is of limited value.

Quality assurance covers the whole process, not just the production, but having assurance on the quality of the production process is a major element (see 10.3.3).

9 Specifications and supervision

9.1 Interpreting specifications

The specification is a means of communicating the requirements of the designer to those involved in turning the design into reality. It should be a definition of the quality the designer requires and is willing to pay for. To get the required quality, it is essential that it is not open to different interpretations; in other words it is objective, rather than subjective.

9.1.1 *Uses of concrete specifications*

In the construction process the specification for concrete, as a material, is subjected to scrutiny by numerous people carrying out widely differing activities, each putting emphasis on different parts of the specification and tending to interpret the requirements to the benefit of the interests they represent (Fig. 9.1)

Activity	*Interpretation*
Designing the structure	Deciding the qualities required of concrete in its finished hardened state and incorporating them in a specification.
Billing	Scheduling the different qualities individually described in the specification and drawings etc, and taken-off volumes of each.
Estimating (1)	Interpreting the different qualities individually described in the bill and any qualities additionally required by the construction process, the volumes of each and translating them into enquiries.
Quoting	Interpreting the qualities described in the enquiry (and the overall requirements of the specification, if it has been provided with the enquiry) and cheaper alternatives available, which still comply with the specification as given in the enquiry.
Estimating (2)	Pricing the concrete as described in the specification/enquiry/quotation.
Buying	Purchasing the concrete as scheduled in the programme (at the best price).

(Contd.)

INTERPRETING THE SPECIFICATION

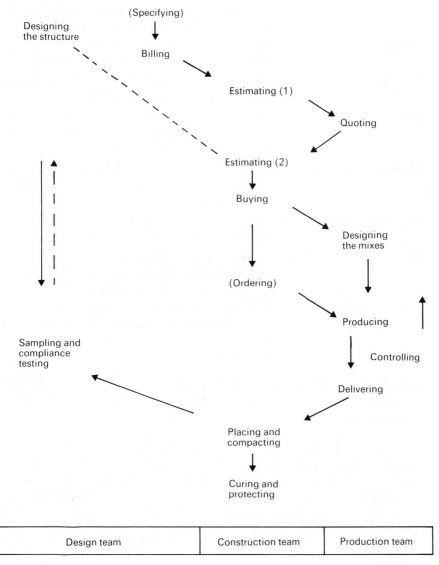

Figure 9.1 Interpreting the specification

Activity	Interpretation
Designing the mixes	Selecting the materials and designing the concrete mix that will meet the purchaser's agreed quality requirements, based on existing data.
Producing the mixes	Mixing the constituent materials in the correct proportions to meet the mix design criteria.
Ordering	Placing the order for the concrete as agreed between the purchaser and supplier
Delivery	Transporting the mixed concrete to site and ensuring that the properties of the fresh concrete are as agreed with the purchaser are met.
Controlling the quality	Testing the mixed concrete to ensure that it meets the mix design criteria and taking action to ensure that it continues to do so.
Compliance testing	Testing the mixed concrete to ensure that it meets the specified quality requirements.
Placing and compacting	Applying skilled workmanship to the fresh concrete to ensure it will meet the specified qualities in the hardened state in the construction.
Curing and protecting	Applying practical workmanship to ensure that the newly hardened concrete will meet the specified quality requirements in the fully hardened state in the structure.

At each activity the wording changes and it is so easy for the potential end quality to change dramatically, depending upon the interpretation and the action taken.

9.1.1.1 *Designing the structure.* The very versatility of concrete makes it an ideal material to solve a wide variety of construction design problems, but care needs to be taken, starting at the design stage, to avoid a few problems that can occur if it is badly specified or misused. It is essential to keep any problems and their solutions, in perspective (Fig. 9.2). Whilst the total solutions to these problems may be highly complex at the design and specification writing stage, a simplistic approach can frequently achieve more than including a treatise on concrete technology in the specification. Tables 9.1, 9.2 and 9.3 provide information on precautions against sulphate, acid and freeze/thaw attack respectively.

(i) Sulphate attack: To combat sulphate attack ensure that the concrete element is substantial enough, the mix is of an adequate grade to provide impermeable concrete and the cement used has sulphate-resisting characteristics: this requires knowledge of soil and groundwater sulphate conditions at the design stage (see 1.2.7).

THE SOLUTIONS

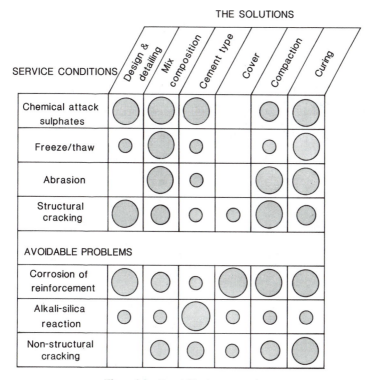

Figure 9.2 Durability in perspective

(ii) Freeze/thaw: entraining air in a concrete of correct grade. If the work is likely to be carried out in cold weather, ensure that adequate facilities for 'winter concreting' are billed (see 1.3.4.).

(iii) Abrasion resistance: specifying a high grade of concrete is an important factor in ensuring an abrasion-resistant concrete, but it needs to be fully compacted and properly cured to develop its potential. (See BS 8110, Tables 3.2, 3.4, Clause 6.5 and Section 6.6).

(iv) Structural cracking: detailing the joints, ensuring enough reinforcement and that an appropriate grade of concrete is fully compacted and properly cured, will minimize the risk of structural cracking. Adequate time should be allowed for before striking the formwork (see 3.2)

(v) Corrosion of reinforcement: Design and detailing should provide for adequate cover to the steel, but it is essential to ensure that it is achieved on site. Cover needs to be physically checked before the concrete is poured, and with a cover-meter after the concrete is cast. If there are likely to be difficulties in achieving the cover required, specify a higher grade of concrete. If there are likely to be difficulties in compacting the concrete around the steel, specify high workability. Specifying the correct nominal size of aggregate to suit the cover is also important (see 3.4).

Table 9.1 Precautions against sulphate attack (see also Table 1.15)

| | Concentrations of sulphate expressed as SO_3 | | | | | |
| | In soil | | | | | |
Class	Total SO_3 (%)	SO_3 in 2:1 water:soil extract g/l	In ground-water g/l	Type of cement	Minimum cement (2) kg/m³	Maximum free water/cement (2) ratio
1	<0.2	<1.0	<0.3	BS 12 BS 1370: lhpc BS 4027: srpc, lasrpc BS 146: pbfc BS 4246: ppfac BS 12 cements combined with: ggbs (4) pfa (5) BS 6588: ppfac BS 6610: pozc	–	–
2	0.2 to 0.5	1.0 to 1.9	0.3 to 1.2	BS 12 BS 146: pbfc BS 6588: ppfac BS 12 cements combined with: <70% ggbs <20% pfa	330	0.50
				BS 12 cements combined with: 70–90% ggbs (4) 25–40% pfa (5)	310	0.55
				BS 4027: srpc	280	0.55
3	0.5 to 1.0	1.9 to 3.1	1.2 to 2.5	BS 12 cements combined with: 70–90% ggbs (4) 25–40% pfa (5)	380	0.45
				BS 4027: srpc	330	0.50
4	1.0 to 2.0	3.1 to 5.6	2.5 to 5.0	BS 4027: srpc (6)	370	0.45
5	>2	>5.6	>5.0	BS 4027: srpc + protective coating	370	0.45

Notes:
(1) This table is based on Table 7 of BS 5328: Part 1: 1991
(2) Inclusive of content of pfa or ggbs. These cement contents relate to 20 mm nominal maximum size aggregate. In order to maintain the cement content of the mortar fraction at similar values, the minimum cement contents given should be increased by 50 kg/m³ for 10 mm nominal size aggregate and may be decreased by 40 kg/m³ for 40 mm nominal maximum size aggregate
(3) When using the strip foundations and trench fill for low-rise buildings to Class 1 sulphate conditions further relaxation to 220 kg/m³ in the cement content for C 20 grade concrete is permissible
(4) Ground granulated blastfurnace slag to BS 6699
(5) Pulverized fuel ash to BS 3892: Part 1: 1982
(6) BRE Digest 363, July 1991 permits ggbs and pfa blends to meet Class 4 condition

Table 9.2 Precautions against acid attack

Environment	Exposure conditions	Groundwater			Requirements for Unreinforced concrete		
		pH	SO_3 g/l	SO_4 mg/l	Min cement content kg/m³ (4)	Max W/C ratio (4)	Lowest strength grade
Moderate	Concrete in contact with non-aggressive soil or concrete surface continuously under non-aggressive water	> 5.5 (3)	< 0.3	< 250	275	0.65	C30
Severe	Concrete surfaces exposed to severe rain				300	0.60	C35
Very severe	Exposed to corrosive fumes				325	0.55	C35 (2)
Extreme	Concrete surface exposed to flowing acid water	⩽ 4.5			350 (Portland cements not recommended)	0.50	C45

Notes:
(1) This table is based on Tables 3.2 and 6.2 of BS 8110 Part 1: 1985. For alternative recommendations see also BRE Digest 363.
(2) Applicable only to air-entrained concrete.
(3) Unprotected Portland cement concrete should not be used in persistent acid conditions of pH 5.5 or less.
(4) Inclusive of ggbs or pfa content.

Table 9.3 Precautions against freeze/thaw attack.

Nominal maximum aggregate size mm	Target air content %
10	7.5
20	5.5
40	4.5

Notes:
(1) This table is based on Clause 4.3.3 of BS 5328: Part 1: 1991
(2) It is applicable to concretes of grade C50 and below.

(vi) Alkali–silica reaction: the maximum alkali level of the concrete or an alternative route permitted in accordance with the Concrete Society report should be specified (see 1.2.8)

(vii) Non-structural cracking: to minimize non-structural cracking requires a cohesive mix which can be readily compacted around the reinforcement. However, the main requirement is proper and sustained curing, which requires strict and continuous supervision (see 3.2).

9.1.1.2 *Specifying the concrete.* An integral part of the design process is specifying the concrete to ensure that the qualities assumed in the design are incorporated in the construction. Guidance on aligning various mix characteristics to grades is given in section 4.6.1.

Determining the type of mix to be specified is an important decision to be made at the specifying stage. *Designated mixes* are quality assured designed mixes linked to particular applications (see 9.2.3.). The characteristics of designated mixes are set out in Table A.1 and typical applications are given in Table A.4. The mix design and control of all the appropriate mix characteristics, not just strength, are audited as part of the QA procedures. *Designed mixes* have been the commonest mixes in use and enable the ready-mixed concrete producer to apply his expertise to best meet the specified requirements with the materials to be used. *Prescribed mixes* need only be used for very special applications, for example when a particular type of exposed aggregate finish is to be produced. *Standard mixes* may be relevant for specifying small quantities of site-mixed concrete, but are not normally appropriate for ready-mixed concrete. Details of standard mixes are given in BS 5328: Part 2 Section 4 and some typical applications are included in Table A.4.

To be assured of the required quality of concrete as a material, it is important that before billing, the following decisions are made:

(i) What grade(s), based on design parameters and site conditions
(ii) What cements, based on site service conditions
(iii) What aggregate size, based on structural section
(iv) Which admixtures, based on site conditions
(v) What compliance rules, based on BS 5328: Part 4 and BS 1881.

9.1.1.3 *Billing.* The scheduling of the various qualities of concrete can raise questions that only those involved in the design can answer, so it is important to appreciate, at the billing stage, the factors that can affect the pricing and have engineering judgements made before commercial decisions are taken. The type of cement is usually the most important cost factor, so it is essential to bill, for example, sulphate-resisting concrete separately to normal concrete. The price difference can be quite considerable, unless ggbs or pfa are permitted as an alternative. Similarly, air-entrained concrete can be different in price from plain concrete of the same grade, and concrete of high workability costs more to produce than concrete of the same grade with medium workability. The specifier should not normally need to state the workability. The constructor should be allowed to select the workability to suit the site conditions.

9.1.1.4 *Estimating (1).* At the time of preparing an estimate, recognition of the different qualities of concrete required by the design is necessary, but just as necessary is an understanding of site practices. Rationalizing grades of concrete used on a particular job can have cost benefits, as well as reducing the chances of confusion on site. High-workability concrete can speed placing, and higher grades can improve stripping times. The time to seek those advantages is before commercial decisions have been taken.

The user of the fresh concrete needs to stipulate the workability required, usually in terms of slump. The traditional 50 mm slump concrete is not suitable for piling or for elements with congested reinforcement, where the degree of compaction it is likely to receive is questionable. It is much better to price on high workability at the start.

9.1.1.5 *Quoting.* Reputable ready-mixed concrete suppliers have a wealth of data on the characteristics of the constituent materials and the wide range of concretes they can produce. It is essential that the full basis of the concrete specification is made available to them before they submit their quotations, as the effect of obscure compliance requirements can greatly affect the quotation. An example of procedures for processing enquiries and orders is set out in Fig. 9.3 and an example of an enquiry/order review form is given in Fig. 9.4.

9.1.1.6 *Estimating (2).* When quotations are received back by the contractor from the supplier, time is frequently at a premium in preparing the final estimate for submission to the specifier and queries raised in the quotation and options offered are not always analysed. The cheapest price tends to dominate.

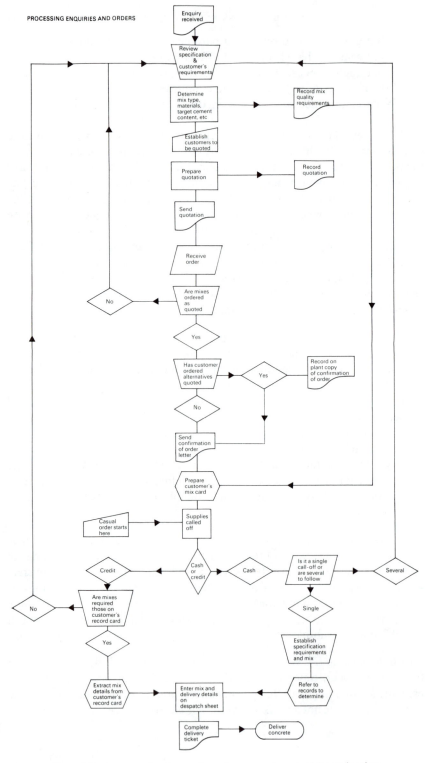

Figure 9.3 An example of procedures for processing enquiries and orders

ENQUIRY/ORDER REVIEW

JOB   Ref No.

Requirement	Detail given with enquiry/order	Notes
1. Basis of Specification		
1.1 Source document		
1.2 Usage of the concrete		
1.3		
2. Service conditions		
2.1 Sulphate condition		
2.2 Exposure condition		
2.3		
2.4		
2.5 Min cover to reinforcement		
3. Workmanship		
3.1 Workability		Check with 1.2, 3.3
3.2 Placing method		
3.3 Curing regime		
4. Materials		
4.1 Cement types		Check with 1, 3.2
4.2 Admixture types		Check with 1.2
4.3 Coarse agg max nom size		Check with 2.5
4.4		
5. Mix designation		
5.1 Max w/c ratio		Check with 2, 3.1
5.2 Min cement content		Check with 2
5.3 Grade		Check with 5.1, 5.2
5.4 Prescription		
6. Limitations		

MIX TO BE QUOTED

Figure 9.4 An enquiry order/review form example

DAILY BATCH RECORD

Plant:

Date: Weather: Air Temperature: ^{o}C – ^{o}C

Ticket No.	Truck No.	Mix	Qty m³	Mix Code	Cmt Type	Slump mm	Admix Type	Cmt 1 kg	Cmt 2 kg	Sand m/c	Added Water l	Loading Time	Remarks

MATERIAL							
Opening Stock							
Deliveries							
Batched							
Closing Stock							
Visual Stock Check							

Figure 9.5   A basic daily batch record

The more specific the specification, the less likelihood there will be of variations.

9.1.1.7. Buying. Those involved in buying have to have a wide knowledge of many materials and processes. The finer points which could affect quality are not always recognized when cost paring is required. Again, the more specific the specification the less likelihood of misunderstandings.

9.1.1.8 Designing the mixes. The options for constituent materials available to make quality concrete with particular characteristics are widening all the time and will increase into the 1990s. It is more important for the engineer to decide on the basic qualities that are required of the concrete, rather than try to fathom out how to achieve them. That expertise lies elsewhere, with the specialist concrete technologist and the experienced concrete producer. With so much data available on the performance of the production concrete, trial mixes rarely need to be called for and this is the guidance given in BS 5328: Part 3: 1990.

9.1.1.9 Ordering. The calling forward of orders by the site from ready-mixed concrete plant/shipping office is a contracted process, and the order should relate to the previous agreement between the purchaser and the supplier. If it is different, then the order would normally be queried by the supplier (see Fig. 9.2). If there was no agreement, then one has to be made at the time of ordering.

9.1.1.10 *Concrete production.* At an average ready-mixed concrete plant operated by one of the major producers, it is normal to stock 10 mm and 20 mm aggregate, one type of sand pc, pfa or ggbs, air-entraining admixture and plasticizing admixture. The other two cements which are occasionally stocked are srpc and rhpc. Lhpc is rarely available, but low-heat qualities are achievable if ggbs or pfa is stocked, as are sulphate-resisting characteristics. Lightweight aggregates are stocked when required, and some plants have 40 mm aggregate, depending upon local availability, and of course, in addition to the two normal admixtures, others can be obtained at short notice. Taking the basic materials, the number of options currently available to concrete purchasers is 810 different designed mixes alone, without considering prescription mixes.

A very basic batch card is shown in Fig. 8.5, but more sophisticated computerized systems are in regular use in many companies. Similarly, the elementary batch record shown in Fig. 9.5 is computerized at many plants and integrates into a complete control system.

9.1.1.11 *Delivering.* The UK ready-mixed concrete industry has settled on the truckmixer with 6 m³ capacity as the standard vehicle, which seems to suit the bulk of the construction demand. Optimizing pour size can lead to more

Figure 9.6 Batching. Courtesy C & G Concrete Ltd

economic construction, but involving the ready-mixed concrete producer in pre-planning can yield many benefits (see section 8.3).

9.1.1.12 *Controlling the quality.* Ready-mixed concrete producers operate quality control systems to ensure that the concrete meets the purchasers' quality requirements. Continuous measurement is necessary to establish variations, trends and the margins required to ensure compliance with the quality criteria, so QSRMC members' laboratories are accredited by NAMAS. Knowing the performance data of the concrete, action can be speedily taken to ensure continuing compliance (see Chapter 6). Stock control plays an important part in the production control process [3].

9.1.1.13 *Compliance testing.* Whilst the water/cement ratio cannot be easily or accurately measured, batching procedures can be observed to check cement content of mixes. The compliance rules set out in BS 5328: Part 4: 1990 provide a practical solution for giving assurance that the concrete supplied complies with the specified grade.

BS 5328: Part 4, Clause 3.5 gives tolerances when using the slump tests, but for some high-workability concretes the flow test as described in BS 1881: Part 105: 1983 may be more appropriate.

Any compliance testing can only be valid if the sampling and preparing of the specimens to be tested has been carried out in compliance with the relevant standards. The 1983 layout of BS 1881 is a considerable improvement in clarifying the testing techniques that should be used, but ensuring that they are complied with goes beyond a few words written into a contract specification. Most test houses are now subjected to accreditation through NAMAS, which can relieve the site control staff of supervising the cube crushing, but the sampling and preparation of specimens on many sites needs continual supervision and insistence on the completion of certificates for sampling, specimen making and testing as required by the Standards. NAMAS has a sampling accreditation scheme.

9.1.1.14 *Placing and compacting* [124]. Placing the supplied concrete should be the easiest part of the business, but inadequate preparation frequently leads to delays which can have repercussions far beyond the site concerned. There is little the specifier can do to protect against inadequate site management, but ensuring the correct size of aggregate is specified and permitting the contractor to select the workability can be a major factor in getting the concrete placed. The designer/specifier could also help to speed the placing process by thinking in terms of pumped concrete, not just for very large pours, prior to billing.

Proper compaction of the concrete is of prime importance if the hardened concrete is to have any chance of performing according to the designer's concepts, so it is important that the correct workability, to suit the construction process, is decided before the concrete is priced.

Figure 9.7 Control laboratory. Courtesy Tilcon Ltd

9.1.1.15 *Curing and protecting* [125]. The importance of curing to ensure durability of the hardened concrete has always been recognized, but rarely accepted on site, because it can impose time constraints, appears to yield no visible benefits, and the requirements are rarely enforced by the site control staff. BS 8110: Part 1, Section 6.6, gives guidance on curing and Table 6.5 indicates minimum curing periods. These curing times may not be sufficient to achieve durability in the very important cover zone or exposed surface, but no matter what cement or cement blend is used, ensuring that concrete is properly cured needs positive site management and specified enforcement which has to be highlighted in the specification.

If the final finish of the concrete is to be exposed, it is usually protected at an early age from physical damage, but concreting in cold weather does, in the UK, need particular attention and an item in the specification should be included to ensure good site practice. Basic curing techniques can only cope with average conditions and special attention does need to be paid to the curing process in hot and cold weather conditions.

9.2 Specifying and supervising the supply of ready-mixed concrete [126]

9.2.1 *Specification clauses*

To ensure that the supply of ready-mixed concrete meets the requirements of the design it is only necessary to specify that:

(i) Ready-mixed concrete shall comply with BS 5328: Part 3: 1990
(ii) Concrete shall comply with the requirements scheduled
(iii) The purchaser of the concrete shall provide all details of the concrete specification to the ready-mixed concrete supplier
(iv) Ready-mixed concrete shall be supplied from a plant certificated by, or meeting the requirements of the Quality Scheme for Ready Mixed Concrete (QSRMC)

MIX DESIGN DETAILS

	Ref: Date:
Materials and mix design information for supply of concrete to:	Plant:

MATERIALS:

Type	Cement Supplier	Works	Aggregates Size & Type	Supplier	Source	Other materials	.....	.....

The source of material given relates to current production.

MIX DETAILS:

	Mix description			Mix design					
Grade	Nom max agg size mm	Cement type	Target slump mm	Mix make–up by dry weight kg. Target Cement Content	Aggregates 1	2	3	4	Max w/c

The mix design details are subject to modification when the properties of the materials vary.

ADDITIONAL INFORMATION:

Signature: Date:

Figure 9.8   Mix design details: a basic format

(v) The purchaser of the concrete shall obtain and submit to the specifier details of the proposed materials and mixes

(vi) The purchaser of the concrete shall, on request, submit to the specifier copies of delivery tickets for ready-mixed concrete

(vii) Where testing of concrete for compliance is required, sampling, making and curing of specimens shall comply with the relevant parts of BS 1881 and BS 5328

(viii) Concrete cube specimens shall be tested by laboratories accredited by NAMAS.

CERTIFICATE OF SAMPLING AND TESTING FRESH CONCRETE		Serial No.
Customer:	Site:	
Sample identity no:	Date:	Time:
Location in works of the concrete:	Ambient temp: °C	Weather:
Location of sampling:	Delivery Ticket No:	
Mix:	Plant:	
Sample carried out in accordance with BS1881:Part 101	Sampler:	Signature:

SLUMP TEST   AIR CONTENT TEST

Determination:	1	2	Air Meter No:		Type:
Slump mm:			Agg Corr Fact:	Calibration date:	
Form of slump:			Operating Press:		
Average Slump:	Slump on ticket:		Place tested:		
Time:	Place slumped:		Time tested:		
Remarks:			Method of compaction:		
			Air Content:		
Details of water addition:			Remarks		
Lapse of time from sampling to test:					
Testing carried out in accordance with BS1881:Part 102			Testing carried out in accordance with BS1881:Part 106		
Tester:			Tester:		
Signature:			Signature:		

Figure 9.9 Example of a form of certificate of sampling and testing

9.2.2 Checking

To ensure that the supply of ready-mixed concrete is being carried out in accordance with the specification, check that:

(i) The purchaser has passed the full concrete specification to the ready-mixed concrete supplier

(ii) The ready-mixed concrete supplier is currently certificated by QSRMC

(iii) Details of the proposed materials and mixes have been obtained and that they comply with the contract specification (see Fig. 9.7)

(iv) Delivery tickets confirm that the concrete supplied complies with the contract specification

(v) Appropriate staff have been trained to sample the concrete, carry out site tests, prepare, cure, store and transport cubes in accordance with BS 1881

(vi) Key aspects of sampling, site testing, cube preparation, curing, testing and reporting are in accordance with BS 1881 (see Figs 9.8, 9.9.)

(vii) The laboratory used for cube testing is currently accredited and the results are reported in accordance with BS 1881 (see Fig. 9.10).

9.2.3 Designated mixes

Specifying designated mixes in accordance with BS 5328: Part 2 Section 5 saves the need to provide many clauses in the contract specification yet

CERTIFICATE OF PREPARING AND CURING CUBES	Serial No.						
Sample identity no:	Site:						
Sampling date: Time:	Certificate of sampling received/not received						
Date cubes cast: Time:	Nominal size of cubes: mm						
Method of compaction: hand/vibration: Type of equipment: no. of strokes/duration:							
Identity mark:							
Age cubes to be tested (days)							
(Laboratory reference):							
Moist air curing: Location: method: period: max/min°c:							
Water curing: time of immersion: time of removal: max/min°C:							
Date cubes taken to testing laboratory:							
Remarks							
Cubes prepared and cured in accordance with BS1881 Parts 108 &111	Name: Signature: Date:						

Figure 9.10 Example of a form of certificate of preparing and curing cubes on site

	CUBE TEST REPORT			Serial No.	
	in accordance with BS1881: Part 111:1983 Part 114:1983 Part 116:1983				

TO:	Site				
	Sample identity No.	Date:	Time:		
Copies to:	Certificate of sampling received/not received				
	Certificate of making and curing received/not received				
	Measured true slump: mm Mix:				
	Measured air content (if applicable)				

Identity mark of cubes							
Laboratory reference							
Date received in Laboratory							
Date cubes made							
Date tested							
Age at test days							
Checked dimensions: nom/meas'd mm							
Shape of specimen							
Conditions on receipt							
Conditions of curing in Laboratory $^oC-^oC$							
Conditions at test							
Method of calculating density							
Mass:as rec'd saturated gm							
Density kg/m^3							
Maximum load at failure kN							
Compressive strength N/mm^2							
Appearance of the concrete							
Type of fracture							

Remarks:

Treatment to remove fins	Codes used to describe conditions of specimen:
Certified that curing in the Laboratory and testing carried out in accordance with BS1881 Parts 111, 114, 116	appearance of concrete:

Person Responsible:	Signature:	Date:

Figure 9.11 Example of a cube test report

ensures automatic quality assurance. By identifying the site conditions and the application for which the concrete is to be used, or the application that most closely resembles it, the corresponding mix designation can then be specified. For non-typical applications, use can be made of Tables A.1 and A.3 to select the appropriate designated mix. It is also necessary to specify whether the concrete is to be unreinforced (U), reinforced (R), reinforced and heated (HR) or prestressed [PS]. If the size of the coarse aggregate has to be different from 20 mm this needs to be specified.

10 Ready-mixed concrete on site

10.1 Choosing ready-mixed concrete

Traditionally, concrete was mixed on site but two conflicting requirements, product quality and product costs, had to be reconciled under the conditions existing on many sites. No two construction jobs are exactly the same so the qualities required of the concrete differ. Experience and knowledge of the characteristics of the constituent materials are required to produce a properly balanced and consistent mix (see Chapter 4).

10.1.1 *Site mixing costs* [127]

To establish the true cost of mixing concrete on site it is necessary to cost:

 (i) The materials for the total volume of concrete that would be required
 (ii) The plant and ancillary equipment that would have to be hired, depreciated etc., and the fuel and maintenance that would be needed
(iii) The labour involved
(iv) The requirements for setting up on site and making good the mixing area on completion.

10.1.1.1 *Materials.* Material costs vary according to the bulk delivered, but in determining the tonnages required it is necessary to calculate the weight of materials required to yield a fully compacted cubic metre of concrete for each mix, say:

Cement (typical)	$275 \, kg/m^3$
Fine aggregate	785
Coarse aggregate	1175

In calculating total quantities, allowance should be made for:

 (i) Dimensions as constructed (compared with those billed)
 (ii) Bulk quantities as delivered in terms of full loads
(iii) Wastage on site.

10.1.1.2 *Plant and equipment.* Plant costs attributable to the cost of the concrete for the whole time the plant is on the site, not just the production time, should take into account the following:

(i) Own plant: capital cost, depreciation, interest, and cost at an estimated utilization percentage
(ii) Hired plant: short-term or long-term hire rates.

Allowance should also be made for road tax, insurance and fuel duty for plant having to be used on the public highway.

The plant and equipment required will vary from job to job, but basic requirements can be taken as mixer, dumper, cement storage (bags and silo), aggregate stockbays and water tank. Allowance also needs to be made for fuel, lubricants, spares as a percentage of the basic rate, maintenance, and disposal of waste.

10.1.1.3 *Labour*. The labour will vary with the type of plant and equipment required and rate of production, but will include drivers, plant operators and labourers. Bonus rates need to be allowed for over and above basic rates, and 'on-costs' need to be included. The cost of site labour or plant required for unloading materials can be a substantial item; management time involved in production and quality control also needs to be accounted for.

10.1.1.4 *Site requirements*. All items which contribute to the production costs of the concrete should be included: transporting plant and equipment to site, erection and return, and site works, water supply, electricity supply, aggregate stockpiles, access for delivery, removal of surplus materials and making good the mixing area. Quality control of the concrete involving site testing and reporting is a necessary 'on-cost', involving testing the aggregates, testing the concrete, mix designs, trial mixes and calibration.

10.1.2 *Ready-mixed v. site mixing supply*

In the case of site mixing, if the plant breaks down or materials are delayed, then all production stops but costs continue to rise, and the maximum output is fixed while the minimum output is uneconomic. The ready-mixed concrete producer on the other hand has the resources and technical expertise to produce a wide range of concrete mixes on demand to meet changes in the site programme and allow operator on-site flexibility in scheduling work. The truckmixers can deliver the concrete to the point of placing, with consequently cleaner sites and no wastage of materials. Costing ready-mixed concrete requires only an enquiry giving all the specification requirements to local producers. It is essential to ensure the concrete will have the correct qualities, and that assurance can be obtained by using a certificated plant.

10.1.3 *Ready-mixed concrete plants on site*

Many ready-mixed concrete companies have transportable plants which can be set up on major contracts (Fig. 10.1). The advantage to the main contractor in using an established ready-mixed concrete producer are: spread of funding and improved cash flow; spread of responsibilities and liabilities; experience in

Figure 10.1   Transportable plant at a power station site. Courtesy Tarmac Topmix Ltd

using the constituent materials; expert concrete technology back-up; improved plant utilization; and quality assurance through QSRMC. There can be disadvantages for the ready-mixed concrete producer if the plant is restricted to site supply; the use of the plant cannot be maximized, and truckmixer fleet costs and arrangements have to be different from normal practice.

10.2 Site preparations for ready-mixed concrete

10.2.1 Site/supply liaison

Establishing a close working relationship between the site and the ready-mixed concrete supplier from the time that a supply of concrete is agreed to the completion of the contract, is of the utmost importance for using ready-mixed concrete efficiently.

Steps to establishing a good site/supply liaison include:

(i) Agreement of the representatives in each organization as to who will be responsible for the overall contract, handling day-to-day requirements and advice, accepting and signing for each delivery of concrete, and providing 'out of hours' telephone contact in the case of an emergency
(ii) Arranging telephone facility on the site
(iii) Arranging for the concrete supplier to attend site meetings.

10.2.2 *Setting up a concrete supply*

In setting up a supply of concrete it is advisable to confirm agreement has been reached with the supplier that:

(i) The concrete mixes and materials comply with the job specification
(ii) The concrete mixes are clearly described and the mix descriptions are clearly understood
(iii) The workability and mixes suit the handling, placing and finishing methods to be used
(iv) Back-up is available if required and that approval includes materials used at the back-up plant
(v) The procedures required by the contract specification for approving materials, mixes and the depot have been completed or are in hand
(vi) The site procedure for approving the start of a concrete pour is agreed and available when required and the production quality is controlled (see Appendix 1)
(vii) The site equipment and trained personnel for sampling and testing concrete are available and satisfy BS 1881 standards
(viii) An approved testing laboratory is employed
(ix) The programme for concrete deliveries has been given to the producer
(x) The details for ordering deliveries have been agreed.

10.2.3 *Programming concrete deliveries*

Realistic planning of deliveries is necessary to ensure smooth concrete service to the site. To provide good service, the supplier requires forward notice of concrete requirements for the site to decide the number of truckmixers and the intervals between loads to deliver concrete at the required rate. This can be provided by:

(i) A provisional programme for the complete contract indicating the weeks or days on which peak demand for concrete will occur
(ii) A weekly up-date of the programme including methods of handling and types of pours
(iii) Daily confirmation of the concrete required for the next day.

Figure 10.2   Truckmixer on site. Courtesy Wimpey Hobbs Ltd

The main site factors to be considered in drawing up a programme are the following.

10.2.3.1 *Handling and placing rate for concrete.* A truckmixer can discharge concrete at approximately $0.5\,\text{m}^3/\text{minute}$ and in most cases it will be methods of site transporting, placing and compacting concrete that will determine the rate at which concrete can be used. The rate of supply depends upon the quick turn-round of truckmixers–25 minutes on site is the maximum which is normally allowed. If a truckmixer is kept for 45 minutes, in a day it will be able to do only four deliveries instead of five. The supplier normally charges for delays on site (see 10.4.4).

10.2.3.2 *Transporting concrete on site with a truckmixer.* Truckmixers are large and, although they have some cross-country capability, they are basically road vehicles. Provision should therefore be made for improving access over soft ground, supporting excavations adjacent to the access, improving traction at steep inclines and manoeuvring, standing and passing spaces.

10.3 Ready-mixed concrete on site

10.3.1 *Delivery ticket*

Immediately before discharging the concrete at the point of delivery, the supplier must provide the purchaser with a pre-printed delivery ticket. It must be checked by site staff to ensure that the load of concrete is the correct one for the pour, and indeed that it is being delivered to the right site or part of the site. The delivery ticket [128] is a very important document, so it needs to be treated with care by all those who handle it: the batcherman, the truckmixer driver and the purchaser's representative on site, the accounts clerk and the clerk of works. It is essential that it is correctly filled in, checked and completed at each stage.

BS 5328 lists the items to be included in the delivery ticket, and the mix and materials delivered must be as described. The delivery ticket needs to include the following:

Name or number of the ready-mixed concrete plant
Serial number of the ticket
Date
Truck number
Name of the purchaser
Name and location of the site
Grade and full description of the concrete, including any additional items that have been specified (e.g. minimum cement content and maximum W/C ratio)
Specified workability
Type of cement and limiting proportions of ggbs or pfa, if specified
Nominal maximum size of aggregate
Type or name of admixture, if included
Quantity of concrete in cubic metres
Time of loading
Space for any additional items that have been specified
Arrival time of the truck
Departure time of the truck
Time of completing the discharge
Water added to meet the specified workability
Extra water added at the request of the purchaser together with the authorizing signature.

Signature of the purchaser or his representative for receipt of the concrete
Safety warning
Conditions of sale.

Separate types of delivery dockets are sometimes used for credit sales and cash sales, to facilitate accounting.

10.3.2 *Addition of water at site*

Concrete is often specified and ordered at a workability that is too low to enable it to be fully compacted under the conditions existing on site—for example, the reinforcement is too congested or the compaction equipment available is inadequate. The truckmixer driver may then come under pressure to make the concrete more workable by retempering the mix. The logic on site is that fully compacted concrete with a higher water/cement ratio is better than having honeycombing which will not carry the load or protect the reinforcement.

To safeguard the purchaser and the supplier, truckmixer drivers must ensure that any extra water added at the request of the purchaser or his representative is signed for on the delivery ticket by that person. If a designed mix is being supplied and more water is added than the mix is designed for, then the prime characteristic of the concrete has been changed.

For retempering of fresh concrete, see section 2.9.

10.3.3 *Safety*

Because of the alkaline nature of cement, prolonged contact with wet mortar or concrete may cause minor skin irritations, or, in extreme cases, even burns. The abrasive effect of the sand in the mix can aggravate the situation, so it is necessary to keep those parts of the body covered that may come into contact with wet concrete.

   (i) Protect the hands at all times with waterproof gloves provided with wristbands
  (ii) Wear a long-sleeved shirt
 (iii) Wear long trousers, ideally with knee pads
 (iv) Should any concrete get into the boots, remove them and wash them thoroughly until clean
  (v) Clothes soaked with wet concrete should be taken off and washed, and the skin washed thoroughly to avoid any irritation
 (vi) Wash wet concrete off the skin immediately. If irritation persists after washing, medical attention should be sought
(vii) If cement or concrete enters the eye, immediately wash it out thoroughly with clean water and seek medical treatment without delay.

MAINTENANCE CHECKS FOR PLANT AND EQUIPMENT: DAILY & WEEKLY

Daily routine   Tick when job completed

1 Adjust tare weights and clean weighing dials

2 Ensure weighing hoppers empty properly

3 Wash out mixer

4 Drain water traps on air lines

5 Shake out silo filters and maintain in efficient working order

6 Ensure truckmixer drums are washed out

Weekly routine

7 Check area under plant for spillage, trade source and rectify

8 Clean yard, ensuring that all drains and traps are clear

9 Maintain settlement pits and wash-down in efficient (and safe) working order

10 Ensure all storage bins and doors are operating efficiently

11 Check conveyors for free-running and wear, adjust as necessary

12 Carry out routine checks and servicing on loading shovel(s)

13 Carry out routine checks and servicing on compressors

14 Maintain all hoppers and doors in clean and efficient working order

15 Shake out cement silo filters and maintain in efficient working order

16 Check dust seals on cement hoppers for wear

17 Check knife edges on weighing equipment

18 Check mixer blades and arms for wear, tightness and clearance; adjust as necessary

19 Remove any cement or concrete build-up in mixer

20 Check calibration of moisture meter

21 Check oil levels on air line lubricators

22 Check water traps on air lines for leaks

23 Check rams and air lines for leaks

24 Check pipework for leaks and wear

25 Check wiring and electrical apparatus for correct operation and report signs of overheating

26 Carry out routine greasing of bearings and gears

27 Carry out routine checks and servicing on mixer

28 Check safety guards are securely in position and walkways clear

29 Report any defects

Batcherman's signature.................... Date...............................

Checked by............................. Date...............................

Figure 10.3 Maintenance checks (daily and weekly) for plant and equipment [128]

<u>MAINTENANCE CHECKS FOR PLANT AND EQUIPMENT: MONTHLY & QUARTERLY</u>

<u>Monthly routine</u>   Tick when job completed

1	Verification of all weigh scales		
2	Verification of water meter		
3	Verification of admixture dispenser		
4	Manufacturer's recommended servicing of loading shovel(s)		
5	Manufacturer's recommended servicing of compressors		
6	Inspect silos		
7	Ensure truckmixers are checked for blade wear		
8	Ensure truckmixer revolution counters are checked		

<u>Quarterly routine</u>

1	Inspection and testing of all weigh scales over their complete range by specialist	
2	Routine oil changes in gear boxes and oil baths	
3	Routine greasing	
4	Ensure truckmixer water meters are verified	

Batcherman's signature....................... Date.....................

Checked by................................. Date.....................

Plant Safety

Ensure at all times, all guard rails and machinery guards are securely fixed in position and walkways kept clean and tidy with clear access.

Figure 10.4   Maintenance checks (monthly and quarterly) for plant and equipment [128]

10.3.4 *Delays*

Ready-mixed concrete is a service as well as a product, so that delays can and do upset shipping schedules. Delays can be classified into three categories: plant, transport and site.

Mechanical breakdown of plants does happen despite planned maintenance, but when it does occur it is often possible to provide concrete from another plant. Daily, weekly, monthly and quarterly check lists are used to

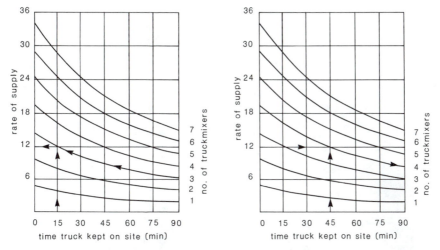

Figure 10.5 Truckmixer deliveries. Truckmixer capacity 6 m³; total journey time 60 min.; load/washout time 15 min. Redrawn and adapted from Newman [141].

ensure routine maintenance is carried out (see Figs 10.3 and 10.4). Restocking of materials can cause delays, particularly when large volumes are ordered at short notice. Early ordering facilitates stock control. With local knowledge, truckmixer drivers are experienced in reducing the effect of traffic delays. Due to the short shelf-life of the product, exemptions from route restrictions are frequently obtained.

The rate of supply of concrete to a site depends on the time the truck is kept on the site. As seen in the example [124] three truckmixers can supply at a rate of 12 m³/h if the time each truck is kept on site is only 15 minutes, but if the time on site is extended to 45 minutes it needs four truckmixers to achieve the same rate of supply.

Truckmixer schedules are upset by delays on site, and the cascade effect can be quite dramatic by the end of a busy day. The need for good communication and understanding between sites and the shipping office is essential if critical pours requiring continuity of supply are not to be affected. In the longer term, truth is more productive than misplaced optimism.

10.3.5 Placing the concrete

Although the truckmixer driver will control the concrete discharging from the truckmixer chute, he can only be expected to do so under the direction of the site staff. In placing the concrete the following principles should be applied:

(i) Deposit the concrete as close as possible to its final position
(ii) Avoid trapping air within the mass of the concrete
(iii) Deposit it in a manner to avoid segregation, in layers rather than heaps

Figure 10.6 Supply to a continuous-flight auger (cfa) piling site. Courtesy Foundation Engineering Ltd

(iv) Fully compact one layer before placing the next one
 (v) The rate of compaction should be related to the rate of discharge
(vi) Avoid the formation of cold joints.

There are many concreting operations which require a continuity of supply to ensure that cold joints are not formed. It is essential that this requirement is made known to the producer at the time of ordering.

Table 10.1 Guide to curing in-situ concrete. [125]

Concrete element	Curing material	Application
1a. Road and paved areas, aprons, open flat slabs	Pigmented resin-based curing compound with high efficiency rating	Apply immediately finishing process has been completed
b. Internal slabs	Polythene or other impervious sheeting material	Protect with shading for the first few hours especially in hot sunshine and high drying winds. Take care to avoid wind tunnel effect under the sheeting.
2a. Tops of beams, columns	Polythene or other impervious sheeting	Apply immediately process has been completed.
b. Top of trench fill, footings, bases		Protect with shading for the first few hours, especially in hot sunshine and high drying winds.
3. Concrete columns beams, walls, etc., which are not to receive subsequent treatment: cast in hot dry conditions	Resin-based curing compound	Apply immediately formwork has been removed.
	Polythene or other impervious sheeting	Fix in close contact with surface immediately formwork has been removed.
	Formwork itself	Leave undisturbed for at least 4 days, preferably 7 days.
4. Concrete columns, beams, walls, etc., where subsequent treatment is envisaged: cast in temperate conditions	Polythene or other impervious sheeting material	Fix in close contact with surface immediately formwork has been removed.
	Formwork itself	Leave undisturbed for at least 4 days preferably 7 days.
5. Formed, permanently exposed concrete sections cast in cold weather	Insulation	Fix as soon as concrete has been placed, and maintain for at least 7 days.
	Delayed removal of formwork	Maintain for at least 7 days.
6. Large concrete sections with a minimum thickness or depth exceeding 1 metre	Top surface insulation	Fix insulation clear of surface as soon as finishing process has been completed.
	Delayed removal of formwork or replacement of formwork by insulating material	Maintain for at least 7 days or until internal temperature gradient is minimized.

10.3.6 *Compacting the concrete*

The ability of the concrete to flow in the formwork around the reinforcement and be fully compacted is a quality that is difficult to define (see 2.2). Within the restrictions of the order, the producer should design a mix that will ensure the concrete is as stable as possible (see section 4.2). The workability of the concrete should be left by the specifier to the contractor/purchaser to determine, based on the method of compaction selected. This avoids difficulties which may arise if the concrete was priced at a lower workability than is actually required to achieve full compaction under the prevailing site conditions.

10.3.7 *Curing the concrete*

Concrete needs to be properly cured to enable it to develop its potential strength and durability, whatever cement is required. The ready-mixed concrete supplier has no responsibility for how the concrete is cured, but there is obvious concern if the concrete which has been well designed and produced is not properly cured. To achieve proper curing requires application of the right curing material [125] (see Table 10.1).

10.4 Pumping concrete [129]

The option of placing the concrete by pump (Figs. 10.7, 10.8) is a very useful one, but to be viable it has to be planned by the contractor. The co-operation between the site staff, the pump hirer and the ready-mixed concrete supplier need to be fully co-ordinated, so that each party is aware that the success of the whole operation depends on the performance of the other two. Large pours are usually well planned and co-ordination is effective. Any pour carried out in an ad-hoc basis usually results in problems for all three parties.

10.4.1 *The contractor*

The role of the contractor can be summarized as follows:

   (i) Co-ordinating the whole operation
  (ii) Establishing the date and time of the pour in consultation with the pump hirer and the concrete supplier
 (iii) Giving adequate notice of the pour
  (iv) Determining maximum output required
   (v) Ascertaining flexibility and reach needed
  (vi) Confirming pumpability of specified mix and obtaining approval of any alterations, prior to ordering
 (vii) Agreeing pipeline layout
(viii) Optimizing pump locations

Figure 10.7 Pumping concrete. Courtesy Tilcon Ltd

(ix) Ensuring access turnround, waiting areas and wash-out facilities for truckmixers

(x) Determining placing and finishing gang size

(xi) Ensuring adequate compacting and finishing equipment (and stand-by equipment)

(xii) Ensuring formwork and reinforcement is ready and checked in time

(xiii) Ensuring formwork and access will withstand pumping vibration and high pressures associated with fast pouring

(xiv) Obtaining approval if necessary of back-up facilities.

10.4.2 *The pump hirer*

The pumping contractor has a responsibility for:

(i) Ensuring the pipelines, if provided by the pumper, has minimum number of and correct types of bends and correct valves fitted

(ii) Advising on optimization of pump locations

(iii) Advising on pump performance in terms of ordered concrete mix

(iv) Earmarking back-up facilities

(v) Ensuring the pump is set up in time to take the first delivery as scheduled.

10.4.3 *The ready-mixed concrete supplier*

The role of the ready-mixed concrete supplier (see section 2.3) can be summarized as follows:

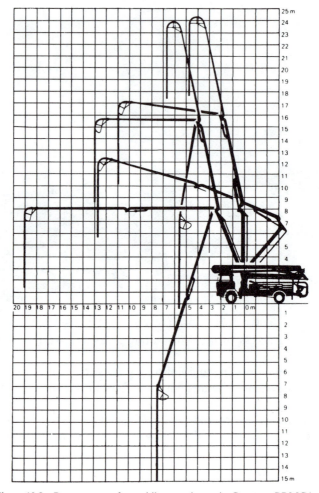

Figure 10.8 Boom range of a mobile pumping unit. Courtesy BRMCA

(i) Advising on the pumpability of mix specified (see section 2.3) with the pump to be used

(ii) Submitting proposals for changes in mix design, if necessary

(iii) Supplying the concrete as ordered

(iv) Advising on date and time of pour to optimize service

(v) Advising on a site arrangements to facilitate access and turnaround of truckmixers.

(vi) Earmarking back-up facilities

(vii) Scheduling concrete supply in conjunction with the pumper and the contractor.

11 Organizations

In the UK there are three organizations particularly concerned with ready-mixed concrete: the British Ready Mixed Concrete Association (BRMCA), British Aggregate Construction Materials Industries (BACMI) and the independent Quality Scheme for Ready Mixed Concrete (QSRMC).

11.1 BRMCA

The British Ready Mixed Concrete Association was formed in 1950 by five companies each operating one plant. As a trade association its role was to represent its members' interests in legal, commercial and technical matters. The need for recognition of the product by the construction industry as a viable alternative to site mixing dominated the work of the association in its early years, and the publication of BS 1926, 'Methods for specifying ready-mixed concrete', in 1962, was a significant milestone in the growth of the industry leading up to 1968 when BRMCA members had over 800 plants in operation. At this stage, the need to ensure that good trade practice was both confirmed and maintained, led to the introduction of the BRMCA's Authorization Scheme. This scheme laid down standards which ensured that plants certified under its rules had the capability to produce quality concrete by employing suitably trained personnel, using materials of proven quality, operating modern plant and equipment and applying established methods of concrete mix design production control. The association employed chartered engineers to inspect the plants and enforce the agreed standards.

In 1972, an important option was added, covering quality control procedures for mixes designed for strength, and by 1981 the scheme covered 1200 registered plants of which 300 were also registered for the designed mix option.

At the end of 1981, with the formation of BACMI, the plants of the ready-mixed concrete industry were divided almost equally between the two trade associations, BRMCA and BACMI. BRMCA continued with its authorization scheme until, in 1984, the independent Quality Assurance Scheme for Ready Mixed Concrete (QSRMC) was formed, when the plants of BRMCA and BACMI came again under a single scheme. Indeed, it became obligatory for members of BRMCA and BACMI for their plants, including those erected on construction sites, to be registered under QSRMC.

At that time BRMCA resumed its original role as a trade association representing its members, particularly in the technical field, but without the direct responsibility for the operation of a quality assurance scheme.

11.2 BACMI

Formed in 1982, British Aggregate Construction Materials Industries (BACMI) is a federation representing the interests in the UK of the producers of construction aggregates (quarried stone, sand, gravel and slag), of lime for industrial and agricultural uses, and of processed materials such as ready-mixed concrete and coated bituminous materials; asphalt and coated-macadam surfacing contractors also form part of the federation.

On the formation of BACMI, the members who had ready-mixed concrete operations formed a group which published a code of practice for the production and delivery of ready-mixed concrete based on BS 5750, Quality systems. It was a self-regulating scheme which was monitored by the federation, but although it fully met the standards of BS 5750 and was technically adequate and very economic to operate, it was not acceptable to many in the construction industry who had come to rely on an inspectorate. In 1984 BACMI supported the establishment of QSRMC, and it is currently mandatory on all members with ready-mixed concrete operations in the UK to register with QSRMC.

BACMI continues to represent all its members' interests, actively lobbying on their behalf and promoting the wide range of products and services which they offer.

11.3 QSRMC

The Quality Scheme for Ready Mixed Concrete is an independent organization solely concerned with quality assessment and certification. The scheme operated by QSRMC provides assurance that concrete is supplied in accordance with quality and quantity requirements of the purchaser. The Scheme employs experienced, professional staff full-time for the assessment of concrete production standards. The Scheme has an independent Governing Board.

11.3.1 *Technical standards* [128]

The technical regulations against which the assessments are made are given in Appendix 1. To be able to give the assurance offered by the Scheme, the producers' operating procedures are assessed against over 400 items in the three categories of order processing, technical records and plant and production.

11.3.2 *Order processing*

To meet the criteria for certification, the producer must demonstrate that his quality system and staff are:

  (i) Interpreting specification requirements correctly
 (ii) Using valid technical data on mixes and materials
(iii) Providing comprehensive quotations

Figure 11.1 Plant inspection by a QSRMC assessor. Courtesy C & G Concrete Ltd and QSRMC

(iv) Receiving orders and issuing correct concrete mix production instructions
(v) Confirming the quality and quantity of materials used against the purchaser's orders.

11.3.3 *Technical records*

To demonstrate that he is capable of achieving the required standards, the producer must show that he has the technical systems and competent personnel able to:

(i) Test materials to British Standards and the purchaser's specification
(ii) Design mixes for performance, prescription, water/cement ratio and yield
(iii) Perform tests in approved laboratories to British Standards
(iv) Use a control system to maintain concrete quality to the required standards.

11.3.4 *Plant and production*

To gain–and maintain–certification, the producer must demonstrate during the assessments and audits that his quality systems and personnel are able to:

(i) Control the delivery, handling, storage and mixing of materials
(ii) Operate batching and mixing equipment
(iii) Comply with batching procedures, instructions and batch recording
(iv) Issue delivery tickets with full mix descriptions.

11.3.5 *Continuing surveillance and enforcement*

After a Certificate of Accreditation has been issued, the producer is subject to:

(i) Regular surveillance visits to confirm the satisfactory operation of the approved quality systems
(ii) Auditing of the quality system data to check compliance with both the QSRMC Regulations and purchasers' requirements.

All certified producers are under a written binding obligation to maintain their standards and to correct any discrepancies promptly. Failure to do so can result in a review, suspension or withdrawal of the Certificate of Accreditation. Continuing compliance with QSRMC Regulations is strictly enforced.

11.3.6 *NACCB accreditation*

In common with other certification bodies, QSRMC is accredited by the National Accreditation Council for Certification Bodies (NACCB). In the case of QSRMC the accreditation is not only for BS 5750: Part 1, Systems approval, but also for product conformity to the new EN 45001 and EN 45012.

11.4 NACCB

The National Accreditation Council for Certification Bodies (NACCB) is responsible for approving Quality Assurance schemes such as CARES (UK Certification Authority for Reinforcing Steels) and QSRMC.

11.5 NAMAS

The National Measurement Accreditation Service (NAMAS), a service of the National Physical Laboratory, is responsible for approval of calibration and testing laboratories. All laboratories are assessed to ensure compliance with the requirements of ISO Guide 25 and EN 45001—the European Standard for the competence of calibration and testing laboratories.

All members of QSRMC have to ensure that their strength testing meets not only the QSRMC regulations, but also NAMAS regulations. Each control laboratory is assessed and accredited, for the appropriate tests, by NAMAS.

Some QSRMC members operate laboratories that are fully accredited by NAMAS for a wider range of tests, and which function as commercial laboratories.

Appendix 1: QSRMC technical regulations

The following schedule forms the basis of the QSRMC Technical Regulations.

1 *Introduction*

Members shall comply with these technical requirements for the Production, Control and Supply of Ready Mixed Concrete.

2 *Quality System*

2.1 An effective and economical quality system shall be established and maintained in accordance with BS 5750 Part 1 to ensure and demonstrate that the concrete conforms to the specification agreed with the purchaser and with these requirements.

2.2 The system shall cover:

* Members' Quality Responsibilities
* Planning to Meet Quality Requirements (Concrete Mixes—Basis of Supply, Description, Materials and Mix Design)
* Production and Delivery
* Materials and Production Control
* Product Quality Control
* Training
* Review of the Quality System
* Records.

2.3 The adoption of the detailed procedures contained in the QSRMC 'Technical Regulations' shall be deemed to satisfy the technical requirements.

2.4 Alternative procedures may be approved if the same degree of assurance is provided, and the procedures are fully documented and available to QSRMC for assessment.

3 *Members' Quality Responsibilities*

3.1 The Member shall nominate one or more representatives to be responsible for implementing and maintaining the standards of the Scheme and shall define the responsibilities and authorities delegated to other staff.

3.2 All staff shall be technically competent for the functions they perform and aware of the effects of these functions on quality.

4   *Technical Definitions*

4.1 Concrete:
Concrete is defined as a mixture of aggregate, cementitious material and water, with or without admixtures and/or additions, and including entrapped or entrained air, for all uses, but excluding all mortars as defined in BS 2787 other than for use as floor and roof screeds.

4.2 Quantity:
The basis of supply shall be the cubic metre of fully compacted, fresh concrete determined in accordance with BS 1881 Part 107. For lean concrete, full compaction is considered to be compaction to refusal in accordance with Clause 1040(3) of the Department of Transport "Specification for Highway Work (1986)". (There is a separate requirement for semi-dry concrete.)

4.3 Designed Mix:
A mix described in performance terms and for which strength testing will form an essential part of judgement of compliance.

4.4 Prescribed Mix:
A mix described in terms of proportions, whether by weight, volume or cement content per cubic metre. A prescribed mix is a mix complying with BS 5328 Part 2 Section 3. Prescribed mixes are mixes selected for particular properties. Strength testing does not form part of the rules for judging compliance of prescribed mixes.

4.5 Cementitious Material:
A cement, or a combination of a cement and pulverised fuel ash complying with BS 3982 Part I, or a combination of a cement and ground granulated blastfurnace slag complying with BS 6699.

4.6 Cement:
Cement is a material defined as such in a British Standard.

4.7 Admixtures:
Admixtures are substances added to mixes in very small quantities (with the exception of pigments typically less than 1% by volume) to modify defined properties of concrete by their chemical and/or physical effect.

4.8 Additions:
Additions are substances, including pulverised fuel ash and ground granulated blastfurnace slag, added to concrete mixes to improve certain properties by their chemical and physical effects and form a significant fraction of the total volume of the concrete.

5   *Planning to Meet Quality Requirements*: (Concrete Mixes—Basis of Supply, Description, Materials and Mix Design)

5.1 Basis of Supply:
  (i) Ready mixed concrete shall be supplied having the quality and in the quantity necessary to satisfy the requirements agreed with the purchaser or his agent.
  (ii) Concrete mixes and materials shall be supplied either:
     a. in accordance with BS 5328 unless agreed otherwise with the purchaser, or
     b. in accordance with the purchaser's specification, or
     c. as agreed alternatives to the purchaser's specification.

5.2 Establishing Quality Requirements:
Specifications and orders shall be systematically reviewed by a competent person to interpret the specified requirements and relate these to mix design criteria. These shall be formally recorded together with any modification to the specification resulting from subsequent documentation to ensure that the plant operator is given the correct instructions for batching and mixing.

5.3 Description of Concrete Mixes:
  (i) Mixes supplied in accordance with Clause 5.1(ii)b shall be agreed in the order and quotation and shall be described as specified.
  (ii) Mixes supplied in accordance with Clause 5.1(ii)a or c shall be described as one or other of the following:
     a. Designed mixes
     b. Designed mixes with additional requirements
     c. Prescribed mixes, either in terms of:
        * special prescribed mixes to BS 5328 using the letters SP and the content of the cementitious material in kilograms per cubic metre (e.g. SP 300).
        * standard mixes
        * proportions by weight or volume
     d. Other descriptions as defined in national specifications.
     e. Mixes supplied as proprietary mixes or marketed under a registered trade name shall be fully described in accordance with Clause 5.3 (ii)a–d and described on the delivery ticket in accordance with Clause 6.5 (i)b.
  (iii) For mixes or materials offered as alternatives to specified mixes or where there is no specification, orders, whether received verbally or in writing, shall be agreed with the purchaser and the fact recorded.

5.4 Constituent Materials
All materials shall be supplied in accordance with the agreement with the purchaser and in accordance with the Regulations.

(i) Cementitious materials

    a. Cementitious materials shall be as specified or any alternative agreed with the purchaser and the fact recorded. The cement type shall be one of the following:

    BS 12, BS 146, BS 4027, BS 6588

    b. Combinations equivalent to BS 146, BS 6588 and BS 4246 may be produced by blending of the constituents in the concrete mixer subject to the provisions of Clause 7.1(ii) and (iii). These combinations shall be described as P/BF, P/PFA and LHP/BF respectively under the heading of cementitious materials.

    d. Combinations equivalent to BS 6588, or BS 6588 Cements shall not be used in Nominal Mixes.

(ii) Aggregates

    a. Aggregates shall be as specified or any alternative agreed with the purchaser and the fact recorded. Where no specification exists, aggregates shall comply with one of the following British Standards:

    BS 882, BS 877, BS 1047, BS 1165, BS 3797

    b. Other aggregates, including types of gradings not covered by one of the above British Standards, may be used by other than ordinary prescribed mixes, provided there are satisfactory data on the properties of concrete made with them.

    c. Separate fine and coarse aggregates shall be used except for BS 5328 grade C15 and below, or whenever concrete is specified by nominal all-in proportions, where all-in reconstituted aggregate may be used.

(iii) Water

Where mains water is not available, water shall comply with Appendix A of BS 3148.

(iv) Admixtures and Additions

    a. For all concrete mixes, admixtures and additions shall only be used where specified or agreed with the purchaser.

    b. Where additions are used, the percentage of addition expressed as a proportion of the cement plus addition content shall be declared to the purchaser.

    c. Pulverized-fuel ash or ground granulated blastfurnace slag may be used as a replacement for any specified minimum cement content, where specified or agreed with the purchaser, the fact recorded and declared in accordance with Clause 6.5(iv).

    d. Any quotation or offer by a Member to include admixtures and additions in a mix shall be made in such a way that the purchaser's agreement is obtained by a definite response to the quotation/offer.

    e. Pulverized-fuel ash shall comply with BS 3892 Part 1 unless otherwise agreed with the purchaser in writing.

    f. Ground granulated blastfurnace slag shall comply with BS 6699.

    h. Pulverised fuel ash or cements to BS 6588 shall not be used as a replacement for any specified cement content in Nominal Mixes.

5.5 Concrete Mix Design
  (i) Concrete mix proportions shall be established for every plant to cover quality and quantity requirements. This programme shall be established prior to production and subsequent to significant changes in materials or concrete properties.
  (ii) All mixes shall be designed to yield the declared volume.

5.6 Designed Mixes:
  (i) Designed mixes shall be designed to achieve a target of not more than 2.5% of test results below the specified strength in order to adequately meet the compliance requirements of BS 5328, unless the purchaser's specification or its compliance rules require a lower value to be adopted.
  (ii) All concrete mix designs shall take full account of the relevance and accuracy of available data relating strength to cement content, the number of test results available from the control system for assessing standard deviation and the achieved rate of testing.
  (iii) The basis of assessment of all concrete mix designs shall be that described in the QSRMC 'Technical Regulations'.

5.7 Prescribed Mixes:
Mix proportions, dependent upon the method of description (see Clause 5.3(ii) c and d), shall be established for each plant supplying prescribed mixes.

6 *Production and Delivery*

6.1 Plant and Equipment:
  (i) At each plant, efficient equipment shall be provided and maintained to ensure the capability to produce and deliver a range of concrete mixes to the standards specified in BS 5328 in respect of quality and quantity
  (ii) Methods of materials storage and handling, and concrete production and delivery, shall ensure that risks are minimised of non-compliance, inter-mingling, contamination, segregation, errors, loss of materials or concrete, and the influences of weather.

6.2 Maintenance:
All plant and equipment shall be maintained in a clean and efficient working condition and regular, routine maintenance checks shall be carried out.

6.3 Measurement of Materials:
  (i) The batching equipment shall be regularly calibrated and maintained to the accuracy of the requirements of BS 5328 and shall permit the plant operative to comply with Clause 6.3(iii).
  (ii) The plant operative shall be provided with a clear display of the quantities of materials to be batched for each mix and batch size and with information

identifying the display to be selected for each designed and prescribed mix to be produced.

(iii) The plant operative shall batch the correct amounts of each material for the required mix design and size of batch within the tolerance of BS 5328.

6.4 Mixing and Transporting:
The methods of mixing and transporting shall comply with BS 5328.

6.5 Delivery Ticket Information:
(i) Immediately before discharging the concrete at the point of delivery, the supplier shall provide the purchaser with a preprinted delivery ticket in accordance with BS 5328.
(ii) The mix shall be described on the delivery ticket in terms of the appropriate category in Clauses 5.3(i) or (ii).
(iii) Combinations of cementitious materials shall be declared in accordance with Clause 5.4(i)b.
(iv) Admixtures and additions shall be declared by word or code which clearly identifies the type or name of the admixture or addition.

7   *Materials and Production Control*

7.1 Control and Purchased Materials Quality:
(i) A control system shall be operated to provide assurance that all materials purchased for, and used in, the production of concrete conform to the standards agreed with the supplier.
(ii) Where combinations of ordinary Portland cement and ground granulated blastfurnace slag are blended in the mixer and declared as P/BF or LHP/BF, tests shall be carried out monthly to confirm equivalence with BS 146 or BS 4246 respectively.
(iii) Where combinations of ordinary Portland cement and pulverised-fuel ash complying with BS 3892 Part 1 are blended in the mixer and declared as P/PFA, tests shall be carried out monthly to confirm equivalence with BS 6588 "Portland pulverised-fuel ash cement".

7.2 Production Control:
The production of concrete at each plant shall be systematically controlled so as to provide assurance that all concrete supplied shall be in accordance with these requirements and with the purchaser's specification, BS 5238 or the agreed description in respect of quality and quantity of concrete.

7.3 Inspection of Concrete:
Each load of mixed concrete shall be inspected following mixing, before despatch and prior to discharge.

7.4 Workability Control:
The workability of concrete shall be controlled on a continuous basis during production and any corrective action necessary taken.

7.5 Additional Water:
No additional water, other than the amount required to produce the specified workability, shall be added to the truckmixer drum before discharge unless specifically requested and signed for by the purchaser.

7.6 Instructions:
Instructions emanating from the operation of the quality system which affect the production control shall be supplied to the production staff.

7.7 Plant Inspections:
Regular routine inspections shall be carried out on the condition of plant and equipment, including delivery vehicles.

8   *Product Quality Control*

8.1 Workability, Plastic Density, Temperature and Air Content
Concrete mixes shall be randomly sampled and tested for workability, plastic density and, where appropriate, temperature and air content. When significant variations from target values are detected, corrective action shall be taken.

8.2 Designed Mixes
   (i) A quality control system shall be operated to control the strength of designed mixes to the levels required by Clause 5.6(i) and shall be based on random tests of mixes which form the major proportion of production. The system shall include continuous analysis of results from cube tests to compare actual with target values together with procedures for modifying mix proportions to correct for observed differences.
   (ii) The control system, when assessed in its entirety, shall ensure that the quality of concrete produced, is at least as high as that required to satisfy Clause 5.6.
   (iii) All upward changes in cement content shall be mandatory unless the cause of the change has been identified and satisfactory corrective action taken. Upward changes shall be made immediately they are indicated as being necessary.
   (iv) Downward changes in cement content indicated by the control system shall be discretionary.

8.3 Prescribed Mixes
Periodic and systematic checks shall be made to ensure that the cement contents of prescribed mixes comply with their mix descriptions.

8.4 Testing Standards and Laboratories:
All testing in the field and laboratory to mix design and quality control purposes shall comply with the appropriate British Standards and, for strength testing, NAMAS regulations.

9   *Training*

9.1 All personnel concerned with production, delivery and the quality system shall have received training appropriate to the duties they perform.

9.2 The testing of materials, proportioning of mixes and the production of concrete together with all its control testing shall be under the overall supervision of an experienced concrete technologist who shall be trained to a standard at least equivalent to the CGLI Concrete Technology and Construction Certificate.

9.3 Concrete shall be batched by operatives who have received proper instructions on the equipment in use and who are able to comply with the required accuracy of batching.

10 *Review of the Quality System*
The quality system established in accordance with these requirements shall be periodically and systematically reviewed at least once each year to ensure continued effectiveness of the system.

11 *Records*
Records shall be maintained to provide continuous assurance of effective operation of the Quality System so as to confirm the quality and quantity of concrete produced. The records shall be retained for the purposes of these requirements for a period of at least one year. The records shall permit traceability and shall include changes occurring, and corrective action taken. They shall cover the following aspects:

  * Planning to Meet Quality Requirements
    — order processing
    — data to substantiate concrete mix designs for quality and quantity
  * Production and Delivery
    — batching instructions
    — batching records
    — delivery tickets
    — equipment calibration and plant maintenance
  * Materials and Production Control
    — concrete production and materials purchase, usage and stocks
    — certificates or test results for materials
  * Product Quality Control
    — control test results
    — quality control system analyses
    — test equipment calibrations
    — laboratory accreditation (where applicable)

Appendix 2: Conversion factors

As calculators are usually available, conversion factors are given rather than tables.

Length	1 in	$= 25.3995$ mm	1 mm	$= 0.03937$ in
	1 ft	$= 0.3048$ m	1 m	$= 3.2808$ ft
	1 yd	$= 0.9144$ m	1 m	$= 1.0936$ yd
	1 mile	$= 1.6093$ km	1 km	$= 0.6214$ mile
Area	1 in^2	$= 645.16$ mm^2	1 mm^2	$= 0.00155$ in^2
	1 ft^2	$= 0.09290$ m^2	1 m^2	$= 10.7639$ ft^2
	1 yd^2	$= 0.83613$ m^2	1 m^2	$= 1.19599$ yd^2
	1 mile^2	$= 2.58999$ km^2	1 km^2	$= 0.3861$ mile^2
Volume	1 in^3	$= 16387.1$ mm^3	1 mm^3	$= 0.0000610$ in^3
	1 ft^3	$= 0.02832$ m^3	1 m^3	$= 35.315$ ft^3
	1 yd^3	$= 0.76455$ m^3	1 m^3	$= 1.30795$ yd^3
Capacity	1 pt	$= 0.56825$ litres	1 litre	$= 1.761$ pt
	1 gal	$= 4.54596$ litres	1 litre	$= 0.220$ gal
Mass	1 oz	$= 28.3495$ g	1 g	$= 0.03527$ oz
	1 lb	$= 0.45359$ kg	1 kg	$= 2.2046$ lb
	1 cwt	$= 50.802$ kg	1 kg	$= 0.01968$ cwt
	1 ton	$= 1016.05$ kg	1000 kg	$= 0.9842$ ton
	1 ton	$= 1.01605$ tonne	1 tonne	$= 0.9842$ ton
Force	1 lbf	$= 0.5359$ kgf	1 kgf	$= 2.2046$ lbf
	1 lbf	$= 4.4482$ N	1 N	$= 0.22481$ lbf
Temperature	1 deg F	$= 0.5556$ deg C	1 deg C	$= 1.8000$ deg F
Velocity	1 ft/sec	$= 0.3048$ m/sec	1 m/sec	$= 3.2808$ ft/sec
	1 mph	$= 1.6093$ km/h	1 km/h	$= 0.6214$ mph
Stress	1 lbf/in^2	$= 0.000703$ kgf/mm^2	1 kgf/mm	$= 1422.3$ lbf/in^2
	1 lbf/in^2	$= 0.006895$ N/mm^2	1 N/mm^2	$= 145.038$ lbf/in^2
Price	1 £/yd^3	$= 1.30795$ £/m^3	1 £/m^3	$= 0.76455$ £/yd^3
Capacity per unit volume	1 pt/yd^3	$= 0.74324$ litres/m^3	1 litre/m^3	$= 1.34551$ pt/yd^3
	1 gal/yd^3	$= 5.9459$ litres/m^3	1 litre/m^3	$= 0.16818$ gal/yd^3
Capacity per mass	1 fl oz/112 lb	$= 27.96.$ ml/50 kg	1 ml/50 kg	$= 0.0358$ fl oz/112 lb
	1 pt/112 lb	$= 0.5593$ litres/50 kg	1 litre/50 kg	$= 1.7880$ pt/112 lb
Fuel consumption	1 mpg	$= 0.340$ km/l	1 km/l	$= 2.825$ mpg
Mass per unit length	1 lb/ft	$= 1.4882$ kg/m	1 kg/m	$= 0.672$ lb/ft
Mass per unit area	1 lb/ft^2	$= 4.8824$ kg/m^2	1 kg/m^2	$= 0.2048$ lb/ft^2
Mass per unit volume (density)	1 lb/in^3	$= 0.02768$ g/mm^3	1 g/mm^3	$= 36.13$ ib/in^3
	1 lb/ft^3	$= 16.0185$ kg/m^3	1 kg/m^3	$= 0.06243$ lb/ft^3
	1 lb/yd^3	$= 0.5933$ kg/m^3	1 kg/m^3	$= 1.6856$ lb/yd^3

Vehicle conversion factors

Tyre pressures								
1bf/in²	20	22	24	26	28	30	32	35
kgf/cm³ (bar)	1.41	1.55	1.69	1.83	1.97	2.11	2.25	2.39
Speeds								
mph	20	30	40	50	60	70	80	90	100
km/h	32	48	64	80	96	112	128	144	160
Consumption								
mpg	5	10	15	20	25	30	35	40
miles/litre	1.1	2.2	3.3	4.4	5.5	6.6	7.7	8.8

Appendix 3: Designated mixes in accordance with BS 5328: Part 2 Section 5

Table A.1 Characteristics of designated mixes with aggregate of 20 mm nominal maximum size

Mix designation	Characteristic strength N/mm^2	Cement group (2)	Min cement content kg/m^3	Max free W/C ratio
GEN 1	10	1	175	N/A
GEN 2	15	1	200	N/A
GEN 3	20	1	220	N/A
GEN 4	25	1	250	0.70
FND 2	35	1 (3)	330	0.50
		2	310	0.55
		3	280	0.55
FND 3	35	2	380	0.45
		3	330	0.50
FND 4 (5)	35	3	370	0.45
PAV 1 (4)	35	1	300	0.60
PAV 2 (4)	40	1(a)	320	0.45
		1(b, c)	340	0.45
RC 30	30	1	275	0.65
RC 35	35	1	300	0.60
RC 40	40	1	325	0.55
RC 45	45	1	350	0.50
RC 50	50	1	400	0.45

Notes:
(1) This table is based on Table 6 of BS 5328: Part 2: 1991
(2) See Table A.2 for cement groups
(3) Depending on the proportion of pfa, a cement complying with BS 6588 may be classified as a Group 2 cement
(4) The PAV mixes contain an air-entraining admixture to give the appropriate air content by volume (see Table 9.3)
(5) See note 2 to Table A.2

Table A.2 Cement groups

Group	Designation	Abbreviation
1(a)	Portland cement complying with BS 12	pc
(b)	Portland pulverised-fuel ash cement complying with BS 6588, or equivalent combinations of pc and pfa complying with BS 3892: Part 1	ppfac
(c)	Portland blastfurnace cement complying with BS 146, or equivalent combinations of pc and ggbs complying with BS 6699—both with 55% ggbs	pbfc
2(a)	Portland pulverised-fuel ash cement complying with BS 6588, or equivalent combinations of pc and pfa complying with BS 3892: Part 1—both with 25% pfa	pfac
(b)	Pozzolanic cement complying with BS 6610, or equivalent combinations of pc with pfa complying with BS 3892: Part 1—both with 40% pfa	ppozc
(c)	Cement complying with BS 4246, or equivalent combinations of pc and ggbs complying with BS 6699—both with 70%. The ggbs shall have an aluminium oxide content of 15%	lhpfc
3	Cement complying with BS 4027 (2)	srpc

Notes:
(1) This table is based on BS 5328: Part 2: 1991, Clause 16.1
(2) BRE Digest 363, July 1991 permits ggbs and pfa cements and combinations with pc to be used in Class 4 sulphate conditions

Table A.3 Equivalent grades for cement content and free W/C ratio for different standard strength classes of cements

Min cement content kg/m³	Max free W/C ratio	Equivalent grades for concretes containing cement classes		
		32.5	37.5 or 42.5	47.5, 52.5, 62.5
200–210	—	C10	C15	C20
220–230	—	C15	C20	C25
240–260	0.70	C20	C25	C30
270–280	0.65	C25	C30	C35
290–310	0.60	C30	C35	C40
320–330	0.55	C35	C40	C45
340–360	0.50	C40	C45	C50
370–390	0.45	C45	C50	C55

Notes:
(1) This table is based on Table 14 of BS 5328: Part 1: 1991 as amended in 1992
(2) The equivalent grade may be selected from the table provided:
    (a) the nominal max aggregate size is between 10 and 40 mm
    (b) the specified slump is in the range 50 mm to 150 mm
    (c) admixtures providing water reduction are not included

Table A.4 Mixes for typical jobs

Job	Designated mix	Standard mix	Slump mm
Blinding	GEN 1	ST1/ST2	75
Concrete fill			
mass	GEN 1	ST1/ST2	75
containing embedded metal	see Reinforced concrete		
Drainage work			
non-aggressive soils	GEN 3	ST4	50
immediate support	GEN 3	ST3	10
Floors			
garage and house (see below)			
wearing surfaces			
general industrial	RC 40	N/A	50
heavy industrial	RC 50	N/A	50
light foot traffic	RC 30	ST4	50
light and trolley traffic	RC 30	ST4	50
Footings			
non-aggressive soils	GEN 3	ST4	75
Foundations			
Class 1 sulphate conditions			
mass concrete	GEN 3	ST4	75
reinforced	RC 35	N/A	75
Class 2 sulphate conditions	FND 2	N/A	75
Class 3 sulphate conditions	FND 3	N/A	75
Class 4 sulphate conditions	FND 4	N/A	75
Trench fill			
non-aggressive soils	GEN 3	ST4	125
Garage floors			
no embedded metal	GEN 4	ST4	75
with embedded metal	see Reinforced concrete		
House floors			
no embedded metal	GEN 4	ST4	75
with embedded metal	see Reinforced concrete		
Housing			
domestic parking	PAV 1	N/A	75
drives	PAV 1	N/A	75
external paving	PAV 1	N/A	75
Industrial floors			
wearing surface			
general industrial	RC 40	N/A	50
heavy industrial	RC 50	N/A	50
Kerb			
backing	GEN 1	ST1	10
bedding	GEN 1	ST1	10
Mass concrete			
fill	GEN 1	ST1/ST2	75
containing embedded metal	see Reinforced concrete		
foundations			
(non-aggressive soils)	GEN 3	ST4	75
Oversite concrete			
below suspended slabs			
non-aggressive soils	GEN 2	ST3	75
Parking			
domestic	PAV 1	N/A	75
Paving			
external	PAV 1	N/A	50
heavy duty	PAV 2	N/A	50

Table A.4 (*Contd.*)

Job	Designated mix	Standard mix	Slump mm
Prestressed concrete			
mild exposure	RC 30	N/A	75
moderate exposure	RC 35	N/A	75
severe exposure	RC 40	N/A	75
most severe exposure	RC 50	N/A	75
Reinforced concrete			
mild exposure	RC 30	N/A	75
moderate exposure	RC 35	N/A	75
severe exposure	RC 40	N/A	75
most severe exposure	RC 50	N/A	75
Strip footings			
non-aggressive soils	GEN 3	ST3	75
Trench fill			
non-aggressive soils	GEN 3	ST4	125
aggressive soils		see Foundations	

Note: This table is based on Table 13 of BS 5328: Part 1: 1991

References

1. Cassel, M. *The Readymixers*. Pencorp, 1986.
2. Wigmore, V.S. Ready-mixed concrete. *Reinforced Concrete Review*, December 1961.
3 Staples, P.N. Trends in ready-mixed concrete. *Concrete*, November 1984.
4. Teychenne, D.C. The use of crushed rock aggregates in concrete. HMSO, Garston 1978, 74 pp.
5. Collis, L. and Fox, R.A. Aggregates: sand, gravel and crushed rock aggregates for construction purposes. The Geological Society, London, 1985, 220 pp.
6. Collins, R.J. Concrete from crushed Jurassic limestone. *Quarry Management and Products*, March 1983, 127–138.
7. Spreull, W.J. and Uren, J.M.L. Marine aggregates and aspects of their use especially in the South East of England. St. Albans Sand and Gravel Co. Ltd., and Civil and Marine Ltd., March 1986, 62 pp.
8. Marine dredged aggregates—a technical appraisal. British Aggregate Construction Materials Industry, London, 1987, 1–15.
9. BACMI Statistical Yearbook, British Aggregate Construction Materials Industry, London, 1990.
10. Carr, M.P. and Banfill, P.F.G. The use of fine marine-dredged sand in concrete. University of Liverpool, September 1984, 70 pp.
11. Banfill, P.F.G. and Carr, M.P. The properties of concrete made with very fine sand. *Concrete*, (London) March 1987, 11–16.
12. Dewar, J.D. The Structure of Fresh Concrete. First Sir Frederick Lea Memorial Lecture, Southampton. Institute of Concrete Technology, 1986, BRMCA reprint, 1–23.
13. Dewar, J.D. Quality requirements of aggregates for ready-mixed concrete. *Quarry Managers' Journal*, London, Vol. 51, No. 9, September 1967, 348–352.
14. Banfill, P.F.G. Alternative materials for concrete—Mersey silt as fine aggregate. *Building and Environment*, 1980, Vol. 15, 181–190.
15. Significance of tests and properties of concrete and concrete making materials. ASTM Special Publication 169B, American Society for Testing and Materials, Philadelphia, 1978, 754.
16. Dewar, J.D. Computerised simulation of aggregate, mortar and concrete mixes. ERMCO Congress, London, 1983, Paper W8B(2).
17. Dewar, J.D. The workability and compressive strength of concrete made with sea water. Cement and Concrete Association TRA/374, December 1963.
18. Russell, P. Effect of aggregates on durability. Supplement to *The Consulting Engineer*, London, April/May 1971.
19. Lees, T.P. Impurities in concreting aggregates. C & CA Guide, Cement and Concrete Association, London, 1987, 7 pp.
20. Dewar, J.D. Effect of mica in the fine aggregate on the water requirement and strength of concrete. Cement and Concrete Association, TRA/370, Slough 1963.
21. Forder, I.E. Some effects of mica in concrete. Ready Mixed Concrete (South West) Ltd., 1971, 34.
22. Concrete Society. Alkali–silica reaction—minimising the risk of damage. 1987. Technical Report No. 30.
23. Shacklock, B.W. Durability of concrete made with sea dredged aggregates. Proc. Symp. Sea Dredged Aggregates for Concrete, SAGA, Slough 1968, 31–33.
24. Chapman, G.P. and Roeder, A.R. Properties of concrete made with aggregates containing sea-shells. SAGA, London, May 1969, SR 6901 34.
25. Grant, N.T. Sands in ready mixed concrete Proc. Symp. Sands for Concrete, SAGA, London 1964, 40–47.

26. Building Research Establishment, BRE Digest 357, Shrinkage of natural aggregates in concrete, HMSO, London, January 1991.

27. Dewar, J.D. Some properties of a lightweight concrete made with expanded perlite aggregate. Cement and Concrete Association, London, TRA/388, May 1965.

28. Concrete Society, Lightweight aggregate for structural concrete. Concrete Society Data Sheet. *Concrete*, London, August 1980.

29. Spratt, B.H. *An Introduction to Lightweight Concrete*. Sixth edn. Cement and Concrete Association, Slough, 1980, 15.

30. Miller, E.W. High density concrete, Parts 1 & 2. Current practice sheet No 90. *Concrete*, London, December 1983, January 1984.

31. Corish, A.T. Personal communication.

32. Corish, A.T. and Jackson, P.J. Portland cement properties—past and present. *Concrete*, London, Vol. 16, No. 7, July 1982, 16–18.

33. ASTM C451-89 Test method for early stiffening of cement paste. American Society for Testing and Materials, Philadelphia.

34. Jackson, P.J. Modern British Portland cements. Conference on Concrete in Construction, Stoneleigh, October 1985, 14.

35. Jones, P.R., Reeves, C.M. and Dewar, J.D. Progress in the use of ground granulated blast furnace slag by the UK ready mixed concrete industry. Int. Conf. on Slags and Blended Cements, Mons, Belgium, September 1981, 10.

36. Dewar, J.D. Quality Assurance of ready mixed concrete incorporating ground granulated blast furnace slag in the United Kingdom. Slag Workshop, Pennsylvania State University, March 1984, 1–13.

37. Dewar, J.D. Standards, codes and quality assurance for the use of ggbfs in ready mixed concrete. Conf. on the Use of ggbfs in Concrete, London, 1986.

38. Dewar, J.D. (1) Introduction to BS 3892: Pulverised fuel ash for use in concrete, (2) Pulverised fuel ash concrete and the UK ready mixed concrete industry and supplement. Joint Seminar: the use of pulverised fuel ash in concrete and in the manufacture of ordinary Portland cement, The Hong Kong Institution of Engineers and The Concrete Society of Hong Kong, September 1982.

39. Dewar, J.D. Specifying composite cements, pulverised-fuel ash and ground granulated blastfurnace slag in concrete *Concrete*, London, August 1985, 30.

40. Bamforth, P.B. Mass concrete. Concrete Society Digest No. 2, Concrete Society, London, 1984, 8.

41. Quality Scheme for Ready Mixed Concrete *Manual of Quality Systems for Concrete*, QSRMC, Walton-on-Thames, May 1984, 31.

42. Basis for certification procedures for testing blends of BS 12 Portland cement and ggbfs or pfa. BRMCA, TB/19/1986, British Ready Mixed Concrete Association, London, 1986.

43. Basis for certification procedures for testing blends of BS 12 Portland cement and ggbfs or pfa. BACMI, Technical Note No. 4, British Aggregate Construction Materials Industry, London, 1987.

44. Building Research Establishment, BRE Digest 363, Sulphate and acid resistance of concrete in the ground, HMSO London, July 1991.

45. Brown, B.V. Air-entrainment–Part 1 and Part 2. Current Practice sheets 80 and 81. *Concrete*, London, December 1982 and January 1983, pp. 59, 60, 45, 46.

46. Waddicor, M.J. A study of some variables within pulverized-fuel ashes which affect the air-entraining ability of admixtures in concrete. ERMCO, London, 1983, Congress paper W12A(3), 7.

47. Dunstan, M.H. High fly ash content concrete. ERMCO, London, 1983, Congress paper W9A(4), 8.

48. Dhir, R.K., Munday, J.G.L. and Ong, L.T. Strength variability of opc/pfa concrete. *Concrete*, June 1981, London, 33–36.

49. Guide to Chemical Admixtures for Concrete. Concrete Society Technical Report No. 18, August 1980, 16.

50. Dewar, J.D. Concrete reasons for admixtures and additions. *Contract Journal*, London, 17 February 1983, 14–15.

51. Sandberg, A. and Collis, L. Toil and troubles on concrete bubbles. *Consulting Engineer*, London, November 1982, 32, 35.

52. Hewlett, P.C. and Edmeades, R.M. Superplasticised concrete, Parts 1 and 2. Current Practice Sheets 94 & 95, *Concrete*, London, April and May 1984, 31, 32.

53. Levitt, M. Pigments for concrete and mortar. *Concrete*, London, March 1985, 21–22.

54. Aitcin, P.C., Pinsonneault, P. and Roy, D.M. Physical and chemical characterization of condensed silica fumes. *Ceramic Bulletin* 63(12), 1984, 1487–1491.

55. Parker, D.G. Microsilica concrete. Part 1: The Material. Part 2: In use. *Concrete*, London, October 1985, 21, 22, March 1986, 19–21.

56. Malhotra, V.M. and Carette, G.G. Silica fume concrete—Properties, applications and limitations. *Concrete International*, May 1983, 40–46.

57. Markestad, S.A. A study of the combined influence of condensed silica fume and a water-reducing admixture on water demand and strength of concrete. Bordas-Gauthier-Villars, Paris, 1986, *Matériaux et Constructions*, Vol. 19, No. 109, 39–47.

58. Page, C.L. Influence of microsilica on compressive strength of concrete made from British cement and aggregates. University of Aston, Birmingham, February 1983, 9.

59. American Concrete Institute, State of the Art Report on fibre re-inforced concrete. Report of ACI Committee 544. *Concrete International*, May 1982, 9–30.

60. Hannant, D.J. Fibre reinforced cement and concrete. Concrete Society, London, February and March 1984. Concrete Current Practice Sheets 92 and 94, pp. 25, 26 and 21, 22.

61. Dewar, J.D. The workability of ready-mixed concrete. RILEM, Leeds 1973.

62. Newman, K. Properties of concrete. *Structural Concrete*, London, Vol. 2, No. 11, September/October 1965, 451–482.

63. Tattersall, G.H. Practical user experience with the two-point workability test. University of Sheffield, May 1983, BS74, 9.

64. Dewar, J.D. Relations between various workability control tests for ready-mixed concrete. Cement and Concrete Association, TRA/375, February 1964.

65. BACMI/BRMCA/CMF/C & CA Joint Working Party 1986: proposal for inclusion in BS 5328.

67. Wiltshire, D.E. Concrete pumping. Concrete Society, Current Practice Sheet No. 79, *Concrete*, London, November 1982.

68. *The Manual of Pumped Concrete Practice*, British Concrete Pumping Association, Harrogate, undated, 1977 or later, 93.

69. Laing Design and Development Centre, Pumping concrete. Digest No. 1. Concrete Society, London, 1984, 7.

70. Gaynor, R.D. Meininger, R.C. and Khan, T.S. Effect of temperature and delivery time on concrete properties. Temperature effects on concrete. ASTM STP 858, American Society for Testing and Materials, Philadelphia, 1985, 68–87.

71. Dewar, J.D. Some effects of prolonged agitation of concrete. Cement and Concrete Association, TRA/367, December 1962.

72. Dewar, J.D. Some effects of prolonged agitation of concrete. London Cement, Lime and Gravel, April 1963, 121–128.

73. Gaynor, R.D. Ready mixed concrete. Significance of tests and properties of concrete and concrete making materials. ASTM Special Publication 169B, American Society for Testing and Materials, Philadelphia, 1978, 471–502.

74. Monks, W. Visual concrete—design and production. *Appearance Matters*, No. 1. Cement and Concrete Association, Slough, 1980, 28.

75. Blake, L.S. Recommendations for the production of high quality concrete surfaces. Cement and Concrete Association, Slough, 1967, 38.

76. Monks, W. Visual concrete—the control of blemishes in concrete. *Appearance Matters*, No. 3, Cement and Concrete Association, Slough, 1981, 20.

77. Non-structural cracks in concrete. Report of a working party. Technical Report No. 22, Concrete Society, London, December 1982, 38.

78. Dewar, J.D. The indirect tensile strength of concrete of high compressive strength. Cement and Concrete Association, TRA/377, March 1964.

79. Dewar, J.D. High strength concrete. Cement and Concrete Association, Slough, 1964, DN/81.

80. Dewar, J.D. Mix design for ready-mixed concrete. *The Municipal Engineer*, London, February 1986, 35–43.

81. Deacon, C. and Dewar, J.D. Concrete durability—specifying more simply and surely by strength. *Concrete*, London, February 1982.
82. Grant, N.T. and Warren, P.A. A CUSUM controlled accelerated curing system for concrete strength forecasting. ERMCO Congress 1977, Stockholm, 314–343.
83. Dhir, R.K. Accelerated strength testing. Concrete Society, London. *Concrete*, October 1976, Current Practice Sheet No. 34.
84. Dhir, R.K. and Gilhespie, R.Y. Concrete quality assessment rapidly and confidently by accelerated strength testing. Congress paper W14A(1), ERMCO, London 1983, 8.
85. Dewar, J.D. Testing concrete for durability. *Concrete*, London, Part 1, Vol. 19, No. 6, June 1985, 40–41; Part 2, Vol. 19, No. 7, July 1985, 40–41.
86. Spears, R.E. The 80 per cent solution to inadequate curing problems. *Concrete International*, April 1983 (quoting research of T.C. Powers and of R.E. Carrier and P.D. Cady).
87. Concrete Mixes—an introduction to the BRMCA Method of Mix Design. British Ready Mixed Concrete Association, Shepperton, May 1984, 21.
88. Teychenne, D.C., Franklin, R.E. and Erntroy, H.C. Design of Normal Concrete Mixes. Department of Environment. HMSO, London, 1975, 31.
89. Deacon, C. and Hopwood, R. National Grades for concrete–progress towards a system of specifying concrete by strength grade in the United Kingdom. Congress paper G8(3), ERMCO, London 1983, 9.
90. Harris, C.A.R. Statistics for concrete—Part 1. Concrete Society Digest No. 5, Concrete Society, London 1984, 7.
91. Barber, P. and Sym, R. An assessment of the variability in ready mixed concrete in the United Kingdom. Congress Paper. W8B(4), ERMCO, London 1983, 1–6.
92. Metcalf, J.B. The specification of concrete strength. Part II: The distribution of strength of concrete for structures in concrete practice. TRRL Report LR 300, Transport and Road Research Laboratory, Crowthorne, 1970, 22.
93. Mathews, D.H. and Metcalf, J.B. The specification of concrete strength. Part III: The design of acceptance criteria for the strength of concrete. TRRL Report LR 301, Transport and Road Research Laboratory, Crowthorne 1970, 18.
94. Metcalf, J.B. The specification of concrete strength. Part I: The statistical implications of some current specifications and codes of practice. TRRL Report LR 299, Transport and Road Research Laboratory, Crowthorne, 1970, 41.
95. Brown, B.V. Statistical compliance with strength specifications. Congress paper W13A(3) ERMCO, London, 1983, 8.
96. Warren, P.A. The Quality Control of Ready Mixed Concrete. RMC Technical Services Limited.
97. Dewar, J.D. Quality control and strength tests—better benefits on way to user. *Construction News*, Supplement, London, 11 May 1972.
98. Dewar, J.D. Specification, quality assurance and quality control of ready mixed concrete. *Municipal Engineer*, June 1987.
99. Day, K.W. Concrete control by cusum method. Congress paper W13A(2), ERMCO, London, 1983, 8.
100. Brown, B.V. Monitoring concrete by the CUSUM system. Concrete Society Digest No. 6, Concrete Society, London, 1984, 8.
101. Sheriff, T. Reference testing and the specification. ERMCO Congress 1977, Stockholm, 269–312.
102. BRMCA Concrete Control System. BRMCA Guide. British Ready Mixed Concrete Association, Shepperton, June 1984, 4.
103. Barber, P.M. and Sym, R. An assessment of a "Target Value" method of quality control. Paper W13A(1), 7th ERMCO Congress, London 22–26 May 1983, 1–7.
104. Pateman, J.D. Influence of site curing on the compressive strength of cubes. *Concrete*, London, February 1977, 30–31.
105. Foote, P. Comparative cube testing—a review. *Concrete*, London, December 1983, C&CA reprint 6/83, 2.
106. Kirkbride, T.W. Testing the testers. Reprint from *Civil Engineering*, December 1975, 3.
107. The performance of existing testing machines. Concrete Society, PCS 62 London, July 1971.
108. Rogers, J. and Hopwood, R. The development of Laboratory accreditation in the United Kingdom. Congress paper W14A(4), ERMCO, London, 1983, 8.

109. Concrete Laboratories: Register of Test Houses. British Ready Mixed Concrete Association, London, published annually, 5.

110. Osbaeck, B. Heat of hydration and strength of slag-Portland cement mixtures. Workshop on blastfurnace slag cements and concretes, York, 1985, Vol. 4, 28.

111. *Concrete Core Testing for Strength*. Second edition, including addendum, Technical Report No. 11, Concrete Society, London, 1987.

112. Changes in the properties of ordinary Portland cement and their effects on concrete. A working party report. Concrete Society, London, 1987 (to be published).

113. Newman, K. The ready mixed concrete producers role in-situ/ndt testing. Conference on Testing of Concrete, Ottawa, 1982.

114. Dewar, J.D. Increasing the quality of concrete in an existing structure. RILEM Symp. on Quality Control in Concrete Structures, Stockholm, June 1979, Vol. 2, 29–36.

115. Samarin, A. Private communication.

116. An assessment of the likely test variations in the analysis of fresh concrete. BRMCA TR10, British Ready Mixed Concrete Association, Shepperton, June 1976, 5.

117. The chemical analysis of hardened concrete. Concrete Society Technical Report No. 32, 1989.

118. Ryle, R. Chemical analysis of hardened concrete. RMC Technical Services Ltd., Egham, Surrey, Technical Report No. 64, January 1970, reprinted February 1973, 9.

119. Bennett, P.E. The functional design of the ready-mixed concrete plant. *Quarry Management and Products*, June 1976.

120. Smit, J. Computerisation for quality concrete. *Concrete*, August 1984.

121. Gilman, F.E. Improvement of management techniques for quality production and service in small and medium companies. *Concrete*, May 1986.

122. Truckmixers. BRMCA Fact Sheet 509. British Ready Mixed Concrete Association, June 1981.

123. Aids to placing ready-mixed concrete. BRMCA Information Sheet 214 (Scotland) British Ready Mixed Concrete Association, June 1980.

124. Blackledge, G.F. Placing and compacting concrete. Concrete Society, Current Practice Sheets No. 61 and 62, 1981.

125. Birt, J.C. Curing concrete. Concrete Society Digest, No. 3, The Concrete Society, London, 1984.

126. Specifying concrete—general aspects. BRMCA Advisory Sheet 501. British Ready Mixed Concrete Association, London, July 1986.

127. Costing concrete site-mixing. BRMCA Information Sheet 212. British Ready Mixed Concrete Association, London, March 1979.

128. Quality Scheme for Ready Mixed Concrete, *Manual of Quality Systems for Concrete*, QSRMC, Walton-on-Thames, 1986.

129. Anson, M. Aston D.E. and Cooke, T.H. The pumping of concrete: a comparison between the UK and West Germany. University of Lancaster.

130. Neville, A.M. *Properties of Concrete*. 3rd edition, Pitman Books, 1981, 150 (Fig. 3.8).

131. Beaufait, F.W. Effects of improper handling of ready-mixed concrete. Proc. Int. Conf. on Advances in Ready Mixed Concrete Technology, Dundee, Sept/Oct 1975, ed. R.K. Dhir, Pergamon, Oxford, 359–366.

132. Ryle, R. The influence of test machines on cylinder splitting strength. Technical Report No. 77, RMC Technical Services Ltd, Egham, 1973.

133. Warren, P.A. Assessing the validity of the cube test result. London Concrete Society Symposium: Engineering judgement on the strength of concrete in structures, February 1975, 20.

134. Popovics, S. New formulae for the prediction of the effect of porosity on concrete strength. *ACI Journal*, March-April 1985, 136–146.

135. Stilwell, J. BRMCA figures show up test house failings. *New Civil Engineer*, London, 31 August 1972.

136. Anderson, R. Ready mixed concrete. *Journal of the Institute of Clerks of Works*, October, November, December 1979.

137. BRMCA Guide: British Ready Mixed Concrete Association, 1971.

138. Gage, M. and Newman, K. *Specification and Use of Ready Mixed Concrete*. The Architectural Press, 1972.

139. Anderson, R. Developments in Works Practice—Concrete Specification. Institute of Works and Highways Technician Engineers.
140. Gaynor, R.D. and Mullarky, J.I. Mixing concrete in a truck mixer. Publication No. 148, NRMCA, 1975.
141. Ready mixed concrete and its role in the construction industry. K. Newman: Proc. Int. Conf. on Advances in Ready-Mixed Concrete Technology Dundee 1975, ed. R.K. Dhir, Pergamon, Oxford.
142. Tattersall, G.H. *The Workability of Concrete.* Viewpoint Publications, Cement and Concrete Association, 1976.
143. Illingworth, J.R. Getting the sums right. Proc. Int. Conf. on Advances in Ready-Mixed Concrete Technology, Dundee 1975. ed. R.K. Dhir, Pergamon, Oxford.
144. Crowther, G. A review of production methods and their developments in the ready-mixed concrete industry. Proc. Int. Conf. on Advances in Ready-Mixed Concrete Technology, Dundee 1975, ed. R.K. Dhir, Pergamon, Oxford.
145. Building Research Establishment, BRE Digest 330. Alkali aggregate reactions in concrete. March 1988.

References to standards

British Standards related to concrete

BS 12   Specification for Portland cement
BS 146   Portland blastfurnace cements
BS 812   Methods for sampling and testing of mineral aggregates, sands and fillers
BS 877   Foamed or expanded blastfurnace slag lightweight aggregate for concrete
BS 882   Aggregates from natural sources for concrete
BS 1014 Pigments for Portland cement and Portland cement products
BS 1047 Specification for air-cooled blastfurnace slag aggregate for concrete
BS 1165 Clinker aggregate for concrete
BS 1305 Batch type concrete mixers
BS 1610 Specification for the grading of the forces applied by materials testing machines
BS 1881 Testing concrete
BS 2787 Glossary of terms for concrete and reinforced concrete
BS 3148 Methods of tests for water for making concrete
BS 3797 Specification for lightweight aggregates for concrete
BS 3892 Pulverized-fuel ash for use in concrete
BS 4027 Specification for sulfate-resisting Portland cement
BS 4246 Low heat Portland blastfurnace cement
BS 4408 Non-destructive methods of test for concrete
BS 4550 Cement methods of test
BS 5075 Concrete admixtures
BS 5328 Concrete
BS 5337 Structural use of concrete for retaining aqueous liquids
BS 5502 Code of practice for design of buildings and structures for agriculture
BS 6089 Guide to the assessment of concrete strength in existing structures
BS 6460 Testing laboratories—accreditation
BS 6588 Portland pulverized-fuel ash cements
BS 6610 Pozzolanic cement with pulverized-fuel ash as the pozzolann
BS 6699 Ground granulated blastfurnace slag
BS 8004 Foundations
BS 8103 Structural design of low rise buildings
BS 8110 Structural use of concrete

References related to Quality Assurance

BS 600   Application of statistical methods to industrial standardisation and quality control
BS 5703 Guide to data analysis and quality control using cusum techniques
BS 5750 Quality systems
BS 7501:1989 (EN 45001) Criteria for operation of test laboratories
BS 7502:1989 (EN 45002) Criteria for assessment of testing laboratories
BS 7503:1989 (EN 45003) Criteria for laboratory accreditation bodies
BS 7511:1989 (EN 45011) Criteria for certification bodies operating product certification
BS 7512:1989 (EN 45012) Criteria for certification bodies operating quality system certification
BS 7514:1989 (EN 45014) Criteria for suppliers' declaration of conformity

Other references

Department of Transport Specification for Highway Work (1986)
ISO Guide 25 General requirements for competence of calibration and testing laboratories

Index